Message
With Love

To

Christian, Lukas and the young plus the future generations

Contents

Note: View e-book pages with 'Figures' in Portrait and 'Equations-Tables' in Landscape

Acknowledgment:

Author is grateful to the experts named for help in art, review, edit and concepts in the Book.

Cover page/Illustrations:

Bot Roda, Artist

Editors/Reviewers:

Murali Krishna Kalluru, Slokas & Review
Udaya Shankar Rao, Sloka & Review
Raghava Rao Mandava, Review
Aruna Koneru, Edit & Review
Anna Scrimenti Edit & Review
Fred Stella, Review
Sivaji R. Vallurupalli, Swastika concept

Preface
god-*Isvar's* Swastika

It is the *sanatana dharma* of a living being to do the righteous duty in striving for purity of conscience and attain the *swastika* of pure *Isvara*.

What is the *dharma* of *swastika* and where is *Isvara*?

Dharma is the law of nature or righteous duty—it is "reality," not a religion. The *swastika* is display of *dharma* by *Isvar*. Before one* can find how *dharma* attains the *swastika* of *Isvar* or look for *Isvara*, one should know what one is looking for. Begin the inquiry perhaps with a simple question: "Who am I and why I am here?"

mah-Isvar or simply *Isvar* in the Brahmi script of *Sanskrit* sound means the Supreme—the Creator/Protector, also known as *Krishna*, the Holy Spirit, or Allah. *Isvara*, a part of *Isvar*, appears within a living being as the *shakti* of *Brahma*, *Vishnu*, *Siva*, and *Narada*. That brings up a new question. Why does *Isvara* appear within a living being in four forms of *shakti*? And if *Isvara* is within a living being, where is *Isvar*?

The simple answer is that the four forms within describe the nature of *Isvara*. One can discover the pure *shakti* of *Isvar* in the *swastika*, observing the precision of the motions of *surya*, *bhoomi*, *chandra*, and the *grahas*. This process is like a detective examining the footprints to look for an invisible magician. The ancients, as the astute detectives, discovered the Almighty *Brahman* or the *param-brahman*, so one can also partake in this discovery, if only one wants to rediscover. How is this possible?

Practice *dharma*, the righteous duty—acquire *jnana*, the intellect of Veda, or *yoga*, unite with the *param-brahman*, or do *bhakti*, prayer. That will lead one to rediscover *mah-Isvar* within the Almighty *Brahman*, and the living beings live within the *param-brahman* with a part of *mah-Isvar*, as *Isvara* or *Atman* or *shakti*, within the *gudi* of a living being on the *bhoomi*.

As the acts of *shakti* dwindle, *Brahma*, *Vishnu*, *Siva*, and *Narada* will cease to be present within the *paramanuvu*, *anuvu*, and *gudi*— the *shakti* to act is gone ending the 'free will' of the senses. The desire for wealth, celebrity status, or power over others, will end too.

The omniscient ancients knew of *param-brahman*, the pure science of Nature. The ancients clearly enunciated the knowledge, about which they had 'heard' or known, in the Vedas for the benefit of the *manu* and for the way of living. That living is to be through *dharma* or righteous duty to purify the conscience within as pure *Isvara*. Thus relieved of the birth-death cycle, pure *Isvara* can join *mah-Isvar* and do *dharma*—directly to the *param-brahman*.

In a *yuga* Cycle of 4,320,000 years, the Veda reveals itself once in the first phase—the pole reversal of *bhoomi* incorporates the *yuga* Cycle. A *yuga* is divided into four phases of unequal time periods. In the present *yuga* we are in the last quarter of the third phase, with over 3,600,000 years passed. The survivors in the polar and equator regions from the Ice Age (IA) of more than 1,000,000 years have begun to multiply, and this outgrowth began about 30,000 years ago. Over the course of the remaining 720,000 years in this *yuga*, the orbiting sun and its companion-star would come onto the same side and go onto the opposite side of *Indra* once in ~1,000 years cycle and will continue going for a total of 720 cycles before the end of time. During each cycle the living beings on Earth will face:

(1) Little hot age (LHA) with rising temperature for a quarter of the time (like at the present) as the sun moves closer to the companion-star on the same side of *Indra*—sun going in an inner orbit (elliptical) and companion going in an outer orbit (elliptical),
(2) A transition from hot to warm age for a quarter of the time as the sun moves away from the companion,
(3) Little ice age (LIA) with falling temperature for a quarter of the time as the sun moves farther away from the companion to the opposite side of *Indra*, and
(4) A transition from ice to warm age for a quarter of the time as the sun moves towards the companion—now, we are at the end of this period, and back in (1).

The above four events continue to repeat in ~1,000-year cycles—like the earth going around the sun in an elliptical orbit in ~365-day cycles with change of the seasons in four quarters. As a general rule in each 1,000 year cycle:

(i) the intensity in LHA should rise and the intensity in LIA should fall in the polar region facing the sun; and

(ii) the intensity in LHA should fall and the intensity in LIA should rise in the equator region and in the polar region not facing the sun—due to increase in the tilt or spin of Earth's axis of rotation in each cycle. In reality, the actual intensity of heat in LHA or of cold in LIA would vary in each cycle as a result of-

- a change in the actual distance between the sun and the companion in each cycle when they arrive on the same or opposite side of *Indra*—due to the orbital paths they follow, and a change in the sunspot intensity of the sun and/or the companion at the time.

The fourth and final phase of the present *yuga*, the end of time with Hot Age (HA), shall begin ~250,000 years from now—per the calculations given in Chapter 9.

During the last Ice Age (IA- with Earth's axis of rotation at 0N or no-tilt), parts of the Veda were lost, which put the man into confusion, and the skin colors drove the man into the darkness of ignorance. This short-version of an ancient history plus the knowledge presented in the form of a dramatic comedy (Chapters 1-8) tells us a story of the confused-Man, who in the dark creates cults in the name of god 'whoever that might be' or claims there is 'no god'. Certain aspects in the comedy drama (Chapter 4 & 6) may sound cynical to some who may be in love with the villain—no matter what harm was done to other life. However, remember a reality-drama requires a Hero and a Villain to make the history of chaos come alive. It is the nature that creates the Hero (*Rishi*) and the Villain (*Rishi*'s disciple, a Pharisee) for a purpose—*Maya* of self v. selfish. The villain, an *asura* with free-will, does things with a selfish mindset as the 'survival of the fittest', like a wild tiger and causes chaos that harms others—native Americans, Africans, SE Asians, poor-Man, and innocent-Animals. Thus, the misguided gains control over the rest of *jiva*—for a short time.

In Chapter 4, the scientists perpetuate the goals of the misguided and practice a religion of 'no god' believing in free-will, and engages in gluttonous lifestyle creating fictitious account of science that our-sun is a 'loner' despite finding 'binary' stars everywhere in the space, and secures approval from Nobel—an organization funded from the profits derived by making and selling dynamite, a precursor to the killer bomb. Now, the Scientists find a new 'religion'—Man is the primary cause, or extremely likely 95%, for the warming on the earth. And claim the man-made concentration of the greenhouse gases (CO_2) increases warming of the atmosphere and ocean, diminishes the snow and ice, and raises global mean sea level. That makes it clear the scientists are either confused or lost—it is the nature, not the man, causing 95% of the rise in temperature on the earth. Man is responsible for the 'garbage' and a small portion of the carbon-dioxide (CO_2)—detrimental to the life on the earth.

The drama in Chapters 1-7 sets the stage on the beliefs of the confused, and the new science and the evidence presented in Chapters 8-9 dispels those beliefs. One can find evidence of the reality of Nature within the *param-brahman* displayed by *mah-Isvar* through the *swastika*, and the acts of *Isvara* within the living being.

The new science confirms that all living beings on Earth live within an Atom of the Almighty-*Brahman*. Thus it is the *dharma* or righteous duty of the living being to seek purity of conscience and attain pure *Isvara*.

avibhaktam cha bhooteshu
vibhaktamiva cha sthitam
bhootabhartru cha tangneyam
grasishNu prabhavishNu cha

Undivided, yet, exists as if divided in beings
Known as the Atman of beings
Devours and generates.
(*Bhagavad Gita*)

*One (Man/Son) meant male and female species. Some of the dialogs in the drama, like Chapter 4, may be in the form of songs....

Glossary: *Sanskrit* words are in *italic*

Sanatan dharma=Eternal law of Nature, the Righteous act

Swastika=Display of *dharma* by the *shakti* of *Isvar* within a *paramanuvu, anuvu* or

Anuvuparamanuvu, anuvu or *Anuvu*=atom or Atom

Shakti=Force or Power of *Isvar*

mah-Isvar or *Isvar*= *Shakti* of the creator/protector

Isvara=A part of *Isvar* as the creator/protector within a living being

Brahma= *Shakti* of Conscience that manifests action

Vishnu= *Shakti* of Conscience in cold that manifests union

Siva= *Shakti* of Conscience in hot that destroys union

Narada= *Shakti* of Conscience that transmits signal to *Vishnu* and *Siva* (like in the thought-controlled bionic leg or computer voice)

Jiva=Life, Self, Soul,

Atma or *Atman*=*Isvara*

param-brahman=Almighty *Brahman*=The Nature or Universe

Manu=Man or Brahman

Veda or *Upanishad*=Knowledge of Almighty-*Brahman* in four Vedas:

Rigveda=Prayers to Nature

Yajurveda=Ritual acts

Atharvaveda=Spells, and

Samaveda=Art, music.

Bhagavad-Gita=*Brahman*-song by *Isvar*

Jnana=Intellect of *Veda*s

Yoga=Union with *param-brahman*

Bhakti=Prayer

Sanskrit=Sound of vibration, the language of Vedas

Brahmi=Pictorial script of Vedas written left to right, the root script of *Manu*

Gudi=Temple or Body

Surya=Sun

Bhoomi=*Bhumi* or *Bhu*=Earth

Chandra=Moon

Grahas=Planets

Yuga=Time lapse between Pole-reversal of the earth

Maya=Mystery

asura=god of senses or free will or selfish

sura=god of righteous or self

Introduction

From: Mikki
Sent: Friday, November 22, 2013 8:22 AM
To: missy.stephens@heritage.org; maggie.delahoyde@heritage.org; leslie.mcclellan@heritage.org
Cc: robert.gordon@heritage.org; jbast@heartland.org; bferguson@sppinstitute.org; nau@gwu.edu; david.davenport@stanford.edu; alkaiser4th@aol.com
Subject: A brief history of God in the United States of America

11/22/13

I cite below from an article *"A brief history of God in the United States of America"* which I read yesterday:

"… the founding fathers… many of them were deists, … deism grew out of The Age of Enlightenment, and that its adherents, while … accept(ing) the existence of a deity, …highly skeptical of organized religion, miracles and the notion that a god intervened directly in the lives of individuals.

The deist philosophy permeates both The Declaration of Independence (in its brief references to God) and the Constitution (in its lack of same). Consider the Declaration (emphasis added):

When in the Course of human events, it becomes necessary for one people to dissolve the political bands which have connected them with another, and to assume among the powers of the earth, the separate and equal station to which **the Laws of Nature and**

of Nature's God entitle them, a decent respect to the opinions of mankind requires that they should declare the causes which impel them to the separation. We hold these truths to be self-evident, that all men are created equal, that they are endowed by **their Creator** with certain unalienable Rights that among these are <u>Life</u>, <u>Liberty</u> and the <u>pursuit</u> of <u>Happiness</u>..."

This one simple concept tells us the whole Truth, nothing but the Truth: "Laws of Nature and of Nature's God entitle...that all men (women?)... are endowed by their Creator with certain unalienable Rights, that among these are Life, Liberty and the pursuit of Happiness".

I think, the Founders were truly 'genius' in recognizing the 'law of nature is the act of god, the creator'.
That is the essence of 'Veda' (ancient knowledge), although the Founders (or we) have not seen the 'evidence' or 'proof' (yet) that the "Nature is the god, the creator".
I have that 'evidence'. And, that is the reason I believe the Founders were truly 'genius'—we can unify all People on this Earth "under god"—I know, nothing-else works.

I wish Dr. Feulner, the Founder of Heritage, would be interested to help move the Founder's simple concept forward and let the young see the 'evidence' that the Founders were truly 'genius'.
I think, that step, a simple step, is an important step and it can be useful to all.

I will be happy to present the 'evidence' to Dr. Feulner and others who might be interested to see.

Thanks,
Maheswar

From: Mikki

Sent: Wednesday, November 20, 2013 12:29 PM

To: 'david.davenport@stanford.edu'; maggie.delahoyde@heritage.org

Cc: 'schieron@stanford.edu'; maggie.delahoyde@heritage.org; leslie.mcclellan@heritage.org; robert.gordon@heritage.org; jbast@heartland.org; bferguson@sppinstitute.org; nau@gwu.edu

Subject: Traditional conservative or conservatism v. new deal

11/19/13

Dear Prof. Davenport and the great founder-Heritage Feulner:

I am the guy who came all the way from India, some 50-years ago, to meet both of you today, and listen to you on the topic of Conservatism or New deal.

It appears as if the topic raps around the concept of 'liberty'-but, the simple concept involves 'free' to worship and do 'righteous duty'. It is like the 'trust and verify'.

That is 'trust in god' and 'verify the man'—is it?

Then, you may ask- what is this 'free' to worship and do 'righteous duty'?

That meant, although 'god' is one (please see the Atom of god in which we live), Man is incapable to see or sense or imagine this One god. However, Man can sense the acts of god.

We should also ask- whose Atom is this in which we live? Or who makes the earth go around itself or go around the sun? Or for that matter- who is running the show within each one of us?

We are told we are made in the image of 'god'. What image is that? Scientists tell us we are made of atoms- Dna, Rna, genes etc.—a key or code in the cell.

If 'god' is made of Atoms, so are we in god's image. So, the key must be the same—but, scientist has no clue as to who turns the key, the gene—is it not?

The one who is turning that key is the 'god': in Sanskrit it is "*Isvar*" and in English it must be the 'holy-spirit' (by the way, English is rooted in Sanskrit).
So, "*Isvar*" must be running the show in the Atom in which we live. That meant a small part of "*Isvar*", within you or I, call it "self" must be keeping us going—for what 'purpose'?
We will find that out later in the Book: god-*Isvar*.

Getting back to liberty or 'free' to worship—why one has to be 'free'?
We can sense the acts of 'god' in a living. Thus, we are 'free' to see 'god' in all the living. So, one must be 'free' to worship anyone that one likes—that liberty is god given for a purpose!

Having said that let me go back to the topic of Conservatism or New deal.
I can't find 'god' in this topic, not directly, anyway. So, this topic must belong to the 'righteous duty' or 'verify'.

So, which one- Conservatism or New deal- is the 'righteous duty'? Which one displays 'god' to 'verify' the man?
Hard to pick—is it not?
Whatever one picks, it only reflects own 'selfish' nature, mostly- as the senses dictate it.
That is the *Maya* (Mystery): self v. selfish

That way, we have two choices: self or selfish. That is the struggle. The Conservatism or New deal is simply a part of selfish nature....

In this context we should also remember the Brahmin or Aryan, the Sura (who worships self, knowing it is the driving force for the righteous duty).

And the Pharisee, the *Asura* (who worships the senses, thinks it is the driving force for the righteous duty).

It appears Asura is ahead, like the rabbit in a race with the tortoise?

In the end, "god-*Isvar*" decides the winner! That is the Judgment Day?

Therefore, we must aim towards the 'righteous duty', not just Conservatism or New deal.

I am an old traditional conservative. That meant 'doing the righteous duty to the best of ability with a view to serve the purpose'.

Yes, I have been a confused-Pharisee for a long time, like most of us…

I am attempting to move away from selfish nature…. I hope god will allow me to do that?

By the way, I am proud of the work being done by Dr. Feulner and others in this struggle: self v. selfish

Please look into a Book: "god-*Isvar*" (Amazon.com, Books- a revised version is expected to be out by end of this month?).

Please feel free to write to me if I can be of any assistance.

Thanks,

Maheswar

Chapter: 1

vedamanoochyaachaaryo antevaasinam anusaasti,
satyamvada, dharmamchara, svaadhyaanmaapramadaha,
svaadhidyapravachanaabhyaam napramaditavyam

Having taught the *Veda*,
Teacher instructs the disciple,
Speak the truth, practice virtue.
Let there be no neglect to study and teach
(*Taittiriya Upanishad*)

Prince learns *dharma*, the righteous duty

The *ashram* is a hut in a hill-valley with woods, and it is on one bank of a river that flows from north to south. Next to the hut stands a large tree with a beehive and bird nests. Monkeys and a variety of small animals use the tree branches as their playing field, and not far from the tree exists an anthill. Apes, deer, snakes, and other animals roam in the bushes nearby. Elephants, tigers, and other wild animals roam freely on the far side of the riverbank. The backyard of the hut is home to a cow, its calf, and a goat. Also in the backyard is a small rice field, a vegetable garden, and fruit trees; a flower garden in front of the hut extends to the large tree. The *ashram* is a small temple-residence for the *rishi* and his disciples.

The *ashram* is in *Bharat*, the land of knowledge, from where the king of *manu* on this *bhoomi* puts *dharma* into practice. Thousands of years ago this king and his queen brought their only son to the *ashram* and studied the Vedas under the guidance of the *rishi* of that time.

The king, queen, and the young prince arrive at the *ashram* in a horse-drawn buggy driven by a soldier. The *rishi* and a disciple receive the guests at the front of the *ashram*.

The king, queen, and prince walk to the front of the *ashram* to meet and greet the *rishi*. "*Namaste, maha-Rishi,*" says the king.

The queen also greets the *rishi*. "*Namaste, maha-Rishi.*"

The *rishi* greets the king and queen and welcomes the guests. "*Namaste, maha-Raja, namaste, maha-Rani,* and *swagatham.*"

The king requests that the *rishi* accept the young prince as a disciple.

The *rishi* agrees and accepts the prince as a disciple.

Fig. 1.1: *Rishi* with the disciple greets king, queen, prince, and soldier

The king, queen, and prince leave their shoes at the door. The *rishi* sprinkles water on the feet of the guests to cleanse their feet. The king, queen, and prince enter the *ashram* led by the *rishi*. They enter a large room where a *murti* with flowers and an eternal flame is present in the northeast corner and ancient maps hang on the wall. Led by the *rishi*, the king, queen, and prince go around the *murti* with a prayer.

Fig.1.2: *Rishi*, king, queen, and prince walk around the *murti* of *Siva*

The king and queen wash the *rishi*'s feet with water and participate in a *puja* that the *rishi* performs, then take *prasadam*.

Then the king looks at the maps on the wall. "I understand these maps, but I am not clear what is going to come next. Is this *yuga*, the life cycle, coming to an end?"

The *rishi* answers the question raised by the king. "No. It is a long way from here; nearly 700,000 years yet to go in this *yuga*. More than 3,600,000 years have passed in this *yuga*. The maps indicate that the *bhoomi* spins to the right until the pole at the top turns to face the *surya*, and that will be the end of this *yuga*. The life on *bhoomi* becomes sparse during the next 250,000 years."

Fig.1.3: *Rishi* and king with the maps on the wall

The king wonders how the *rishi* knows this.

The *rishi* informs the king, "We know the *bhoomi* rotates and orbits, but we do not know how it spins as it rotates around itself once a day-night of twenty-four hours with its rotation speed decreasing each year. It orbits around the *surya* once a year, or in 365 day-nights, with its orbit speed varying with time. *Bhoomi* spins around itself once in two *yuga*s with its spin speed varying with time each year."

The prince asks curiously, "What makes the *bhoomi* spin?"

The *rishi* answers, "The cause is *Indra*."

The prince continues with his inquiry. "Where is *Indra*?"

The *rishi*, holding a *mala* (chain) of beads in his hand, tells him, "This *japamala* has 108 beads. I pray in the names of *Isvar* as I move the *mala* in my fingers, one bead at a time, 108 times in a cycle. The number 108 represents the common factor for the distance between the controller and controlee in an *Anuvu* or *anuvu* or *paramanuvu*. In other words, this 108 relates to the distance between the controller and controlee in terms of the size or diameter of the controller. For example, the distance between the *surya*, the controller, and the *bhoomi*, the controlee, is 108 times the size or diameter of the *surya*.

Fig. 1.4: *Rishi* answers the questions posed by the prince

Likewise, the distance between *Indra*, the controller, and the *surya*, the controlee, is 108 times the size of *Indra*. So each bead in this *japa-mala* represents 1/108th of the distance between the controller and controlee. Or, if you take the distance between the *surya* and the *bhoomi* equal to one bead, this *mala* with 108 beads represents the distance between *Indra* and the *surya*. In other words, the distance between *Indra* and the *surya* equals 108 times the distance between the *surya* and *bhoomi*. And *Indra* is invisible or insensible to the *manu*."

"How do I know this 108 is accurate if *Indra* is insensible?" the prince wonders aloud.

The *rishi* points out the fact that the ancient *rishi*s knew. "That knowledge is a part of the Veda that you can verify."

The prince admits that he lacks the skill to verify this.

"You should first learn about the motions of the *bhoomi* and *surya*," says the *rishi*, "and then learn about the *shakti* of *Indra* and how it works."

"What is *shakti*?" asks the prince.

The *rishi* explains that *shakti* is *Isvar* and creates the *surya*, *bhoomi*, *grahas*, and the rest of the living beings on *bhoomi*. "*Shakti* makes all the living beings act and do their duties," he says.

The prince inquires about free will. "Does the 'free will' make *shakti* do what it wants?"

The *rishi* informs the prince, "You use free will only to discriminate right from wrong and then do your righteous duty."

The prince wonders aloud whether *Isvar* creates each living being to do a specific duty.

The *rishi* assures the prince that this is the reality.

The prince becomes curious and asks, "Is it a duty of *manu* to walk around a *murti*? If it is, what purpose does it serve?"

"The *murti* represents the *shakti* that makes the *surya* and *bhoomi* go around *Indra* performing a duty. The same way the *manu* should go around the *murti* under the *dharma*."

The prince does not understand. "What is this *dharma*?"

The *rishi* defines *dharma* as "the law of nature set by *Isvar* that living beings should follow for purity of conscience."

"What is the law of *Isvar*?" the prince asks.

"The law of *Isvar* appears in four forms of *shakti*: *Brahma*-consciousness, the *Narada*-consciousness signal, *Vishnu*, and *Siva*."

"What are *Vishnu* and *Siva*?" the prince asks.

The *rishi* explains that *Isvar* manifests *Vishnu* as cold *shakti*, which creates a *paramanuvu* union into an *anuvu*, and *Isvar* manifests *Siva* as hot *shakti* that destroys the *anuvu* to make the constituents or *paramanuvu* act or dance. "And the duty being performed by the *surya*, *bhoomi*, and *grahas* in going around *Indra* help mix the cold *anuvu* and hot *paramanuvu*, thus aiding *Vishnu* and *Siva* in doing their duty," he adds.

That makes the prince curious. "Where is *Isvar*?"

"*mah-Isvar* is known as *Isvar*, and a small part of *Isvar* is known as *Isvara*. And, *Isvara* is omnipresent in all living beings as *atma*. *Isvara* is present in *paramanuvu*, *anuvu*, *Indra*, *surya*, *grahas*, *bhoomi*, and in the tree, you, me…in all the living beings…as the self or life."

The prince asks, "What is *atma* or self?"

The *rishi* explains, "*atma* is self or life, and *atma* is *Isvara*."

"Then *Isvara* is the *shakti* of life?" asks the prince.

The *rishi* nods. "That is correct."

Then the prince raises the big question for which no *manu* has ever figured out a clear answer: "What is the purpose of life?"

The *rishi* gives a simple answer. "The purpose of life is to perform happily one's righteous duty and attain *moksha*."

That makes the prince want to know what *moksha* is.

The *rishi* informs the prince that *moksha* means relief for *atman* from the birth-death cycle. It unites with *mah-Isvar* to serve the *param-brahman* directly.

Suddenly they hear a loud noise and a cry for help coming from the river.

Fig.1.5: *Rishi* and king at the rear-door looking out

8

The king readies his sword to go and help, but the *rishi* stops him. "O King, you need not act. The situation is under control; watch the outcome."

A deer, taking reprieve in the river, is now finds itself in the mouth of a hungry crocodile. An intervention from a hippopotamus saves the deer from death. Assisting the deer to safety, the hippopotamus licks the wounds of the deer to heal. After a while, the deer stands up and walks away toward the *Ashram* with a limping foot. The Hippopotamus keeps an eye on the deer satisfied with the outcome.

The *rishi* teaches the moral of the story as doing "righteous duty."

Fig.1.6: Deer, Crocodile and Hippopotamus in the river

While the *rishi*, the king, queen, and prince are inside the *ashram*, the disciple and a soldier are standing near the horse-drawn buggy getting to know each other by exchanging views. At the end of their conversation, the disciple concludes, "I hope we can act with free will for a change."

The soldier agrees. "I am on your team and am ready to act with your blessings."

The king, queen, prince, and the *rishi* come out of the *ashram*. The soldier goes to get the buggy ready, and the disciple joins the *rishi*.

Fig.1.7: Soldier and disciple at horse-drawn buggy scheming the future

The king expresses his gratitude to the *rishi*. "*Dhanyawad, maha-Rishi*. I will always remember the lesson learned from the hippopotamus doing its duty, the righteous duty."

The queen thanks the *rishi*. "*Dhanyawad, maha-Rishi*. I am grateful for your acceptance of my son as a disciple. I am leaving my son with you to learn righteous duty."

The *rishi* blesses them all. Then the king and queen depart with the soldier in the horse-drawn buggy.

Fig.1.8: *Rishi* and disciples with the deer

The *rishi* and his disciples meet the deer in the backyard standing near the cow, calf, and the goat. They tend to the wounds, rubbing them with the medicinal leaves from a nearby bush. To console and make the deer feel secure, they give her water and food and she becomes a new member at the *ashram.*

Glossary:

Ashram =Residence of a *Sage* or *Rishi* (to live with family and disciples)
Namaste=Salutation, I bow before *Isvara* within you
maha-Rishi=Supreme-scholar that leads an esthetic life teaching Vedas and *dharma* of *Isvar*
maha-Raja=Supreme ruler of *Manu*=Man on the Earth, and follower of *dharma*
maha-Rani=Queen of the supreme ruler; *Bharat*=Land of knowledge
Swagatham=Greetings, Welcome
murti=Deity of *Isvar* (*Vishnu*+*Siva*), *Puja*=The ritual of worship to *Isvar*
Prasadam=Fruits or food offered to *Isvar* at the *puja*
Surya=Sun, *Bhoomi*=*Bhu*=Earth, 24 hours=1440 minutes or 86400 seconds

Yuga=4,320,000 years

Indra=Neutron with DNA with *shakti* of *Isvar* (*Vishnu+Siva*), the chief constituent in an Atom

Atma or *Atman, Jiva*, Self, Life or Soul=*Isvara*

Moksha=Relief from birth-death cycle or emancipation from taking rebirth

Dhanyawad=Thanks, *Deva*=God, *Devi*=Goddess

Akash=Sky or Space

Grahas=Planets

Horse-drawn buggy=Passenger carrier drawn by horses

Chapter: 2

Om! dhyouha shantirantariksham shantihi
pruthvee shantiraapaha shantihi
oshadhayashantihi vanaspatayashantihi
viswadevaahashantihi brahma shantihi
sarvam shantihi shantideva shantihi
saamaa shantirodhi Om! shantihi |||

O' Lord of Peace! May there be peace in heaven and on the Earth
May all the forces and the law of the Universe be benevolent to all
(*Yajur Veda*)

Prince learns Veda, the knowledge

The next morning, the *rishi* and his disciples perform their daily routine. They clean inside and outside of the *ashram*. They tend to the needs of the deer, goat, cow, and calf, and they milk the cow and do other work.

As the earth rotates, the red sun appears to be rising in the east. The *rishi* and the disciples take a bath in the river water with a prayer to *Surya Deva*.

Then, as they walk back toward the *ashram*, the prince asks, "Why do we pray to *Surya*?"

Fig.2.1: *Rishi* and disciples bathe in the river with a prayer to Sun god

The *rishi* explains, "The *Surya Deva* is our Father and makes the heat and light for the survival of living beings. The heat turns water in the seas into clouds, and the clouds rain down as water. The light gives us vision, among other things. Thus Father keeps the living beings thriving on the *bhoomi*."

"Why do we worship the *bhoomi*?" the prince inquires.

"The *Bhu Devi* is our Mother and gives us the food and water for the nourishment of all living beings."

"Then who gives us the air to breathe?"

The *rishi* explains, "We primarily breathe the *paramanuvu*, which is eternal. The *paramanuvu* in *akash* gets modified by the cold *shakti* from *Indra* and the hot *shakti* from the *Surya*, and the *bhoomi* mixes it for the use of all living beings. The living beings on the *bhoomi* inhale cold-warm *paramanuvu* and exhale warm-hot *paramanuvu*. That includes you, me, the tree, *bhoomi*, and *surya*."

Then the prince starts to wonder what this *paramanuvu* means.

The *rishi* defines it as an infinitesimal *anuvu* that comes in infinite sizes.

The prince asks, "How did living beings come into existence?"

The *rishi* explains to the prince the fundamental processes involved. "In our universe, all living beings come into existence by the acts of *Indra*, the *Surya*, and the *bhoomi*. When conditions on the *bhoomi* become conducive, living beings get created, like the worms that give birth in the soil and come out on the surface."

"The process simply goes like this: *Isvar* creates *anuvu* using the cold *shakti* of *Vishnu* that comes from *Indra*, and creates *paramanuvu* using the hot *shakti* of *Siva* that comes from the *Surya*. *Bhoomi* attracts that mixed *anuvu* plus *paramanuvu* from *akash* that rains down in all places on the *bhoomi*. It comes down into the swamps as well. At that time, if the conditions are conducive in a swamp, *Isvar* makes egg cells from certain *anuvu* and *paramanuvu* and sperm cells from certain other *anuvu* and *paramanuvu*. When *Isvar* lets a small part of *shakti*, as *Isvara*, to get purified that *Isvara* allows the sperm to join with the egg, the swamp takes the role of a mother's womb in feeding more *anuvu* and *paramanuvu* to the fertilized egg. The *atman* in the fertilized egg accepts certain *anuvu* and *paramanuvu* to grow per the embedded code of instructions set out by *Isvara*."

"The *shakti* of *atman* within the fertilized egg makes the *anuvu* multiply and the cells grow into a living being of a certain type of species with its own unique ability. At the right time, that living being goes out of the swamp into the water or onto the land and/or into *akash* to continue living on its own. Whether the created self will live in water or on land and/or *akash* depends upon the type of specie. The specie that survives will grow, mature, and perform a specific duty. Then the rest of the population in each type multiplies in water, on land, and/or in *akash*."

The prince has doubts about what happens if the sperm and egg do not join.

The *rishi* explains, "The live sperm and live egg, being half-*shakti* each, both give up *shakti* after a short life and the *anuvu* become waste."

Thus the prince understands that to grow into a living being the two half-*shakti*(s) must join.

The *rishi* agrees with the understanding of the prince and goes further to explain it clearly. "Names like *Lakshmi-Vishnu*, *Parvati-Siva*, *Sita-Rama*, and *Radha-Krishna* consist of both female and male names of the goddess-god. Thus the name represents one *shakti* joined by two halves together. That means a half-*shakti* cannot exist alone."

The prince arrives at the conclusion that this means he is an assembly of *anuvu* and *paramanuvu* by one *shakti* of *Isvara*.

The *rishi* agrees with the conclusion of the prince. "Yes, the *shakti* or *atman* that resides within the *paramanuvu* cloud of the fertilized egg drives the embedded code of instructions and makes the code operational in assembling the required *anuvu* into a living being, like you. Your thoughts, actions, ability, duty, appearance, and lifestyle are set by the *atman* (*atma*) based on your prior *karma*.

To help you in that process, your senses assist you in eating, drinking, breathing, and all other activities. The *shakti* within you accepts what is useful and rejects the rest as waste. When the senses stop working properly, the *shakti* becomes weaker and the *anuvu* in the body begins to destroy itself like the sperm and egg cells when the half-*shakti* leaves. Then *Isvar* takes the *paramanuvu* from the sperm/egg or the body cells and reassembles into a new *paramanuvu*, giving the half-*shakti* of life to cause rebirth."

The *rishi* and the disciples enter the *ashram*.

Fig.2.2: *Rishi* and disciples perform *puja*

The *rishi* gives the *murti* a bath in milk while the disciples go out to pick flowers, fruits, and vegetables for presenting to the presiding deity at the *puja*. The disciples wash the items in water and set them up near the *murti*.

Then the *rishi* and his disciples walk around the *murti* with a prayer. The disciples wash the *rishi*'s feet in water and participate in the *puja* performed by the *rishi*.

At the conclusion of the *puja*, the disciples bow before the *murti* with a *namaste*, touch the *rishi*'s feet, receive *prasadam*, and leave the room.

Later, the *rishi* and the disciples sit under the tree, where the prince learns to meditate under the guidance of the *rishi*, and they practice deep breathing while the animals watch.

The prince wonders aloud, "If this is how to meditate, what else is done in *yoga*?"

The *rishi* explains to the prince when to meditate or when to do *yoga*. "You meditate to solve a simple problem. You do that by concentrating your mind calmly on one thing at a time and thinking of a solution. In *yoga*, you meditate on a much bigger problem, like seeing or sensing the reality in the mind. That requires practice to master the body, senses, and mind. So in *yoga*, you meditate until your consciousness unites with the consciousness of the *param-brahman*."

Fig.2.3: *Rishi* teaches deep breathing to the disciples

The prince recalls the *rishi*'s prior message that *moksha* is to serve the *param-brahman* directly. "What is the *param-brahman*?" he asks.

"*param-brahman* is the nature in whom we live."

"If the *param-brahman* is nature, why unite consciousness with the *param-brahman*?" the prince asks.

The *rishi* gives a reason to satisfy the curiosity of the prince: "By the unity of your consciousness with the *param-brahman* you can sense reality and attain bliss."

The prince is confused. "I want to know about this bliss."

The *rishi* puts it in simple words. "Bliss means to feel secure and happy."

"Are all living beings capable of achieving bliss?"

The *rishi* says, "It is possible, but it requires certain ability to achieve it."

The prince asks, "Who might have that ability?"

The *rishi* replies, "All living beings with consciousness will have the ability, but the ability differs between any two individuals. That is the reason the *manu* is divided into *varna*. Some will have the ability to study the Vedas and teach *dharma*; some will have the ability to follow and implement *dharma*; some will have the ability to follow *dharma* in service; and others will have the ability to follow *dharma* in work. Thus each one can contribute to the best of their ability under *dharma*."

The prince wonders aloud, "Who divides the *manu* into a *varna* of ability?"

The *rishi* informs the prince that it is the *Isvar.*

The prince wonders, "How do we know that to be true?"

The *rishi* explains the reality. "You can see in reality that the true nature between any two creatures differs. Look at bees or ants, for example. You find a queen, advisors, managers, and workers. Each one does its own duty to the extent of its ability."

Then the prince moves into another aspect of nature. "Why does the ability differ?"

The *rishi* says simply, "That is the result of prior *karma.*"

The prince is confused and fails to understand how prior *karma* comes into play.

The *rishi* explains, "*Karma* defines the level of purity of one's conscience. An individual's ability to perform a duty depends upon consciousness."

"Does that mean one's ability depends upon the purity of one's conscience?"

"Yes, the level of purity sets the level of ability in performing a duty. And it depends whether or not it is a righteous duty."

The prince asks, "Then what is a righteous duty?"

The *rishi* puts it in simple words. "Live and let live or act as if you and the rest are all One."

The prince thinks back to the bee and ant and wonders if the bee or ant has consciousness.

"Every living being has consciousness to enable it to do a duty," the *rishi* states clearly. "The bee and ant too are created from *paramanuvu*, within which lies the consciousness."

"So each creature performs a specific duty," the prince says.

The *rishi* agrees. "Each one is created to perform a specific duty, but the duty in the next birth differs from the duty in this birth."

"Why does the next birth have a differing duty?"

The *rishi* explains that clearly. "The birth-death cycle continues until the conscience attains purity. The level of purity of conscience in the next birth will differ as it depends upon the righteous or unrighteous acts done in this birth. Righteous acts make the conscience purer and improve its ability; unrighteous acts diminish its ability. When the conscience attains purity, the self becomes free of rebirth, unites with the consciousness of the *param-brahman*, and attains *moksha*."

The prince is not sure what this means. "Does that mean the king or queen, the rich or poor worker, the queen bee or bee or ant all have the same chance to get a more pure conscience in this birth?"

The *rishi* clearly affirms this reality with a yes.

The prince is not clear about the Veda.

The *rishi* defines the *Sanskrit* word *veda* as the "knowledge of nature or *param-brahman*."

The prince is curious as to who discovered that knowledge.

"The knowledge was discovered by an ancient *rishi* a long time ago, during the first phase of this *yuga*," the *rishi* informs the prince. "A *yuga* is 4,320,000 years with four phases of unequal time periods."

The prince asks, "How did the knowledge come to us if that was discovered so long ago?"

The *rishi* explains, "During the last ice age the ancient writings were destroyed. Therefore, some of that ancient knowledge came to us through the word of mouth from our ancients, who memorized the knowledge in the form of the sounds of vibration. Later, other ancients who heard those sounds as *mantras* transcribed the *Sanskrit* sounds onto '*TaaLapatra*' in old Brahmi script."

The prince becomes curious and asks, "How does the old Brahmi script of *Sanskrit* differ from the present script?"

The *rishi* clarifies it this way: "The old Brahmi is written from left to right, and a number of scripts took birth from it that are written from left to right or right to left or up and down. The *Sanskrit* sounds written in old Brahmi can be read or understood by only a very few."

The session concludes. The *rishi* and his disciples continue to meditate under the tree. At the end, the prince bows with a *namaste* before the *rishi* and withdraws into the hut. The other disciple bows with a *namaste* before the *rishi* and remains standing to say good-bye.

The *rishi* knows his disciple's mind and encourages him to speak. "Go ahead. You can choose your own path to follow."

The disciple addresses the *rishi*, saying, "*maha Rishi*, I beg for your mercy. Allow me to submit my departing prayer to you. I learned Veda under your guidance. I am grateful to you for teaching the knowledge. However, I am unable to see any good in *murti* worship or dividing *manu* into *varna*. For that reason, I must go on with a free will in search of my god, whoever that might be."

The *rishi* gives the blessings with advice: "Study and teach Veda, and follow *dharma*. Do not become a Pharisee to Veda."

The disciple bows and touches the *rishi*'s feet, goes into the *ashram*, and comes out with a bag and departs.

Fig.2.4: *Rishi* and the disciple

Glossary:

Puja=Worship to *Isvar*

Karma=Fate, duty assigned to a living being

Mantra=Sound of vibration, a set of syllables with embedded metaphysical power

TaaLapatra=Writings on special-Bark or Leaf or Papyrus,

Brahmi=Ancient pictorial script written left to right

Varna=Ability of a living being to do a righteous duty

Pharisee=Separated from the Varna of Veda

Chapter: 3

pratamaas tapa eva, dvitiyo brahmachaary
achaarya kula vaasee, tritiyo tydantam
aatmaanam achaarya kule vasaadayan
sarve ete punya lokaa bhavanti!

Austerity indeed is the first,
Pursuit of sacred wisdom is second,
And to dwell in the house of teacher is the third.
All these attain the worlds of virtue.
(*Chhaandogya Upanishad*)

Prince is *Buddha*

Some years later, the prince weds in the palace.

The *rishi* is present at the wedding ceremony of the prince.

The prince visits with the *rishi* in a private room after the ceremony and expresses his gratitude. "*Namaste, maha-Rishi*. My stay at the *ashram*, and your teachings of Veda and *dharma*, made me learn, think, and act righteously. I am forever grateful to you. I have no interest in this kingdom or the palace or the riches. I must seek the Truth."

Fig.3.1: Prince and bride walk around a fire at their wedding ceremony

The *rishi* gives his blessings to the prince. "I bless you, and I pray to *Isvar* to grant your wish. Be a *buddha* in this *yuga*."

The king and queen enter the room to thank the *rishi* for vising with them and to say good-bye.

Fig.3.2: *Rishi* blessing Prince after the wedding

Glossary:

Buddha=Learned or enlightened *Yogi*

Chapter: 4

sreyascha preyascha manushyaam
etas tau samparitya vivinakti dheeraaha
sreyohi dheerobhi preyaso vrineete
preyo mando yoga kshemaad vrineete

Both, good and pleasant, approach a man
The wise man chooses the good
The simple minded prefers the pleasant
(*Katha Upanishad*)

God is nowhere?

The high school students enact a drama at the end of the class year. Students in costumes act to entertain their fellow students, teachers, and parents. The main actors are the pope, Isaac Newton, Albert Einstein, and other famous scientists.

The speeches below sound wild to make a point, and the style is in the form of song and dance.

Scene 1:
Copernicus (enters and sits in a chair):
I hope you remember me. I am Copernicus.

The pope (enters next and sits in a chair):
I am the pope, you know me.

Then Kepler (enters and sits next to the pope).

And Galileo (enters and sits next to Johannes Kepler).

Copernicus:
I am Copernicus. I am the one who first saw the planets go around
the sun. Maybe someone else knew that, but I saw it with my own
eyes!

The pope:
I heard that, Copernicus. I am the pope.
Remember, I am the sole authority here in this reality.
So until I say yes, no one can see anything new.

Immediately a voice:
O Pope, wait a minute.
Did you miss seeing Mohammad?
I am here with a sword in hand.
I know you are not the sole authority, Allah is!

Kepler:
I saw the planet Mars.
It goes around the sun in an elliptical orbit.
Do I know why the planet moves that way?
No, I do not. I have no clue.

Galileo:
I saw that Earth is not the center of our universe.
The sun is the center of our universe.

The pope:
Earth is the center of the universe.
And Earth is flat.
O Mohammad, forget your Allah.
The son of god whispered the truth in my ear.
And I believe that is the truth!

Background voice:

Allahu Akbar!

I am Mohammad. I must leave now.

Do I need to go for another jihad?

I will see you, Pope, at the jihad!

Another background voice:

I am the Pharisee and no one can ever forget me!

I am the one who devised the 'right to left' script from Brahmi.

I am the one who devised the Avesta from the Vedas using my free will!

Did I stop there? No, I did not.

I went on to devise the Torah, the Bible…and I did it all.

Of course, in the name of my 'god,' whoever that might be!

And I gave help to my disciples in devising the Koran and Red Book.

Name it, and I am into all the adventures.

If you ever question why I did all this,

The simple answer is I did not believe in the Vedas.

Ask me why?

The Vedas did not recognize me as the Chosen one.

I wish I knew who chose me to do all this work!

If you think I am confused or lost, ask my ex-disciple Mohammad.

O Mohammad, I hope you remember me, your teacher.

I recall you did a good job on the *sura*s, the Veda people.

I can see it from here, the whole drama…

My people enlisted a powerful army, the Christians…

Now my people are poised to take over the earth,

They are more powerful on the earth.

That I can clearly see from up here!

So stop your threat of jihad and surrender as a good soldier.

Immediately, a background voice:

I am Hitler, the unifier under the *swastika*, like Caesar before.

O Pharisee, the divider, you know we met before.

Again, we met in hell, can you remember?

So stop your games or I will be back!

Fig.4.1: Copernicus, Kepler look-on as the police drag Galileo away

Galileo:

I am Galileo, and I know the sun is the center of our universe.

The pope:

Galileo, listen to me and stop that nonsense.

Or you will be in jail the rest of your life.

The pope leaves the scene.

And Galileo falls off of his chair.

The police drag Galileo away. And later the police announce that
Galileo is no more.

Fig.4.2: Newton and Christian debate

Scene 2:

Isaac Newton (enters):

I am Newton, and I saw the gravitational force of the sun.

It makes planets orbit around the sun.

Do I know how the force transfers between the sun and the earth?

No, I do not. I have no clue.

Or what causes the elliptical orbits?

No, I do not. I am clueless.

But I saw light in seven colors.

How did I see that?

I saw when a light ray dispersed through a glass prism.

So I conclude that the white light is made up of seven colors.

A young man (enters):

I am a student with questions, lots of questions.

I am Christian. May I make a comment, sir?

Is it possible the atoms in the glass prism display the colors…

And that the light ray has no colors?

Newton:

Young man, it is possible.

When a light ray hits a tree leaf, what do you see?

You see only one color, green or whatever…

A young woman (enters):

I am Esther, and I wonder, where did the rest of the six colors go?

And a green leaf or green grass turns brown in winter. Why?

Newton:

The atoms assembled by nature display a specific color.

The color green or brown…

It depends upon the leaf's ability to absorb/reflect light.

So the skin color of an animal in cold turns brown…

And in hot it turns darker.

The atoms assembled by man display more colors.

Christian:

Sir, may I suggest…

A green leaf turns brown losing some of its atoms.

A brown skin turns darker losing some of its atoms.

What role does the angle of the incident ray play in all this?

What role does the prism angle plays in all this?

Does the angle have any effect?

Newton:

Certainly the loss of atoms makes a difference.

When a sun ray hits a raindrop at a certain angle, the rainbow appears.

Christian:

What does that mean, sir?

Does that mean light has no color?

Or does the light display the color of an atom when it hits at an angle?

Newton:
Yes, yes. I know it is a mystery!

Christian:
So the light in space is the color of the atoms in space, is it not?

Newton:
It is reasonable to think so.
But space is an empty vacuum.

Christian:
How can an empty space transfer force and make Earth orbit the Sun?

Newton:
I do not know.

Christian:
May I suggest…
After all, the space is not empty!

Newton:
Can you tell me why?

Christian:
I think the reason is simple.
And I wish it were that simple.
I see space as red in the morning with the rising sun.
I see space as red in the evening with the setting sun.
I see space as white or blue the rest of the time with the sun shining.

Newton:
Well said, young man.
In the morning, it is the shining sun at an angle to the clouds.
In the evening, it is the shining sun at an angle to the clouds.

Like the shining sun at an angle to the raindrops showing a rainbow.
And it is the shining sun with no angle at other times.

Christian:
Why do we see only the red in the morning and evening?
Why not seven colors as in a rainbow?

Newton:
I do not know.

Scene 3:
Niels Bohr (enters):
I am Bohr, and I just saw an atom doing its work.
It has one proton and one electron, nothing else.
And I saw the electron going around the proton.
So I confirm my finding; it is the atom H, hydrogen.

A voice:
It is great to see. Here is a Nobel!

Esther:
I have a question, sir.
What is this proton or electron?
What makes the electron go around the proton?

Bohr:
I wish I knew.
I do not know what is going on.
It appears gravity is the cause in the hydrogen atom…
Just as the gravity of Sun on Earth.

Esther:
Chadwick discovered a neutron in the hydrogen atom.
Did you see the neutron?
Do you know what this neutron is doing?

Bohr:
No, I did not see a neutron.
Nor do I know what Chadwick saw.
No, I do not know what is going on.
My fellow scientists assume the neutron holds on to the proton.

Esther:
I wonder why the neutron holds on to the proton.

Bohr:
I have no clue.

Scene 4:
Charles Darwin (enters):
Here I come, the one and only Darwin.
Sorry I am late; I have come directly from Africa.
I had a small talk with our cousins, the apes.
Here is my final report on my discovery.
Man evolved from an ape, no offense.
Our cousins confirm that as a truth.

Esther:
Sir, do I look like an ape?

Darwin:
I am too old to see anymore.
I am not able to see your features well.
Tell me, where did you come from?

Esther:
Not out of Africa.

Darwin:
Are you confusing me?
Or you are like me, not able to see the features?

A young man (enter):
I am Mohammad…
May I ask you a simple question?

Darwin:
Wait a minute.
Are you the jihad Mohammad?

Mohammad:
No sir. I am the kid Mohammad.
You just missed the jihad Mohammad and the pope.
They left to finish an unfinished job!

Darwin:
I am glad to know that you are the kid Mohammad.
Go ahead and ask the question.
I may or may not be able to answer.
I prefer not to offend anyone.

Mohammad:
This is my question, sir.
Is there something in Africa that can prove man evolved from an ape?

Darwin:
I like the question.
My simple answer is look for an ape bone.
I am sure a bone can be found in Africa to provide the evidence.
Sooner or later someone will show up with a bone to prove it.

Scene 5:
Karl Marx (enters):
I am Marx. I see injustice and no god.
The false consciousness of the ruling elite will suffer.
The ruling elite will find the social order as not predetermined.

The nature of the worker is to act in self-interest.
The world transforms in favor of the class struggle.

Mohammad:
What do you mean?
Do you mean give up the Torah, Bible, and Koran?
Take up the Red Book?

Marx:
Yes, I hope you do that.

Scene 6:
George Lamaître (enters):
I am Lamaitre. I saw the big bang at the birth of the universe.
And I know we all came out of this big bang.

Esther:
Sir, I am confused.
Did we come out of Africa or out of the big bang?

Lamaître:
I can understand your confusion.
First I became an expert in the Bible.
Then I became an expert in physics.
That confused me too.
Let me clarify what is going on.
The universe came out of a big bang 13.7 billion years ago.
We came out of Africa some six thousand years ago.
Pharisee saw that event and wrote in the Bible for us to know.

Esther:
You saved my day, sir, and I thank you.

Edwin Hubble (enters):
I am Hubble, a lawyer, and I am the expert on space.

I saw the relative speeds of galaxies, and the universe is expanding.
I admit I do not know what is going on in reality.

Albert Einstein (enters along with his friends):
You all know, I am Einstein.
I am the one who saw a photon!

A voice:
It is great to see. Here is a Nobel.

Einstein:
Thanks, I accept.
I always thought the universe is at a standstill.
A great blunder on my part.
Thanks, Hubble, for alerting me in a timely fashion on your findings.
The lawyer in you is the cause of that confusion.
You said it well, without knowing what is going on in reality.
I can fix your confusion with my equations of space-time.
I can do that with a stroke of my pen, sitting at my desk.
My equations allow for the expansion of the Universe.
Also allow for acceleration of the Universe in the future.
It is all simple magic and I am good at it.
I know how to add or delete a magic constant in general relativity, GR.
I must admit, Lamaître and Hubble, you are geniuses.
Newton, you too must be a genius to see a fictional gravity.
I am the one who knew all about gravity and the orbits of the planets.
I saw the curvature of space-time in the universe sitting in my room.
That curvature is the cause of gravity, orbits, and other phenomena.
Who would have known without me?

Professor Dean Falk (enters):
Here I come. I am Falk, an expert anthropologist in the gulf coast.
Oh Einstein, your brain can do magic—do you know why?
Your brain lobes are complicated convolutions.

How do I know that?

When you were asleep, I went into your brain.

I took photographs of 240 sections and looked at 2,000 slices.

Your brain with unusual folds is average in size and weighs 1,230 grams.

The pattern of ridges in your brain gives that ability to do the magic.

Mohammad:

Professor Einstein, how can you be so sure your equations do the magic?

Einstein:

You must believe in Einstein—didn't you listen to Falk?

Ask other experts, like Schwarzschild, de Sitter, and Eddington.

They can tell you all about Einstein seeing the curvature of space-time.

Karl Schwarzschild (stands up):

I am Schwarzschild, the founding expert on black holes.

I saw a black hole at the center of the galaxy.

I did that using Einstein's GR equations.

Einstein too saw the black hole and approved of my work.

Then Einstein presented my work to the academy.

All the experts on the earth fell for it, no questions raised.

Willem de Sitter (stands up):

I am de Sitter. I tested Einstein's GR equations on lunar orbit.

I proved that the earth and the sun control the motions of the moon.

I know ancients saw an invisible hand moving the objects in space.

I do not believe in that invisible hand—I may be ignorant.

I know Einstein's magic works.

I overrule any need for an invisible hand.

Arthur Eddington (stands up):

I am Eddington. I am a professor at Cambridge…

And I am an official at the Royal Astronomical Society, UK.

I am the one who went into the sun…
And I measured the interior temperature.
It is about 10x10^6K.
And I saw the fusion of the hydrogen, H, into helium, He.
My buddy de Sitter asked me to look at Einstein's GR equations.
I did, and I fell in love with Einstein and his set of complex equations.
I figured out a simple experiment to prove the validity of GR.
I went out of my way to help a fellow scientist in Germany.
At that time the UK and Germany were at war—is that World War II?
I forget counting the wars—so many of them.
I devised a simple scheme to prove the validity of the GR equations.
One way to prove it is to measure the deflected light ray of a star.
Take a photo of a distant star at the time of the sun's eclipse.
Take another photo of the same distant star a day or a month later.
I did that to the best of my ability and compared the two photos.
I figured out the deflection of the star's light ray from the photos.
I calculated a deflection using Einstein's equations.
The result is in agreement, roughly.

Esther:
Did you really go into the sun to measure?

Eddington:
Please don't take it literally…
I am an expert in daydreaming with mathematics.
I am one of those who sit at a desk and solve the problem.

Esther:
Sir, may I ask another question?
I assume you took the photos accurately.
The star was at position A when you took the first photo.
The star moved to position B when you took the second photo.

The earth would move during this entire period.

The star and the earth move in different ways.

How can you measure the deflection of a star's light rays accurately?

Eddington:

I agree.

That's the reason I figured that out to the best of my ability.

I never claimed I measured it accurately—I figured it out roughly.

Mohammad:

May I ask Professor Einstein a follow-up question?

Can you show me the curvature in space that you saw?

Explain what the curvature does to a planet.

Einstein:

I am sure I can.

I heard NASA spent about fifty years and one billion dollars.

And NASA saw the curvature in space near Earth.

If NASA can see it, you can too.

Let us do this simple experiment—

Take this blanket and spread it out tightly between two tables.

Now tie the four edges firmly onto the tables.

Thus the blanket stays suspended in space.

Then place this large ball in the middle of the blanket.

Let us call it the sun.

You see a big dip in the blanket.

Next, let us place two small balls to represent Earth.

Place one ball to the right of the sun at more than a meter away.

Place the other ball to the left of the sun at less than a meter away.

Now you see two small dips and one large dip in the blanket.

These dips define the curvature in space.

The curvature of the large dip controls the motion of the balls.

Mohammad:
Why did we place two balls to represent one Earth?

Einstein:
True, it is only one Earth.
The two balls represent Earth's extreme positions around the sun.

Christian:
May I ask, sir, what is space-time?

Einstein:
You have seen the curvature of space part.
Now let me show you the time part.
Time defines the variation of the curvature in space.

Christian:
How is that possible?

Fig.4.3: Einstein and Mohammad experimenting on curvature of space

Einstein:

I can explain.

Earth will take some time to move from one extreme point to the other.

Earth and its dip move due to the curvature in space created by the sun.

The movement of Earth with its dip depends upon time.

Hence I call it the curvature of space-time.

Mohammad:

I am not sure what that means.

If I let this small ball go free, where does it go?

Does it move around the larger ball and reach the other extreme point?

Einstein:

Yes, let it go free…and see.

Test it.

Mohammad:

I let the small ball go free.

It went straight, hit the large ball, and then stopped.

What went wrong?

Einstein:

This is only a model.

There is no guarantee that a model always works.

The reality is unpredictable.

Mohammad:

What is the reality?

Einstein:

I can say loud and clear…

My GR equations define reality precisely.

Christian:
Please tell me what makes Earth come closer to the sun on one side?
What makes it go farther away from the sun on the other side?

Einstein:
I thought it was the space-time.
Now I am not sure what is going on.
I know no one else does either.

Christian:
I learned that the plane of Earth's orbit is inclined.
What is it and why?

Einstein:
I can answer that.
See the two balls on either side of the sun…
And look at the horizontal plane through the center of the sun.
The ball on the left of the sun is above that plane.
The ball on the right of the sun is below that plane.
That way, the plane of Earth's orbit is inclined to the horizontal plane.
I do not know the reason why.

Christian:
What makes the outer moons of Jupiter and Saturn orbit at high inclinations?

Einstein:
I have no clue.

Christian:
Do you know if the curvature of space-time has anything to do with it?

Einstein:
I do not know.
I did not try to solve that issue.

Christian:
I read that scientists have measured the speed of light on Earth.
Is that true, and what is the speed?

Einstein:
The speed of light is approximately $3 \times 10^8 m/s$.

Christian:
Is the speed a constant between any two points in our universe?

Einstein:
I think the answer is no.
The speed of light varies with space-time.
And the speed of the earth's orbit varies too.

Christian:
May I ask you a question on your invention?
Please tell me how your formula can convert rock or dirt into energy.

Einstein:
My formula is simple.
It says energy equals mass times the speed of light squared or
$E=mc^2$.

Christian:
Does the light come with a mass?

Einstein:
Light has no mass.

Christian:
That means light cannot create energy.

Einstein:
I agree.

Christian:
Does the light acquire speed due to the energy of mass conversion?

Einstein:
Yes, I agree.

Christian:
What makes the speed of light vary?

Einstein:
I think it is entropy.

Christian:
What is entropy?

Einstein:
It means a change in temperature.

Christian:
I have a follow-up question.
Does an increase in temperature at a point increase the speed of light?

Einstein:
My answer is, sure it can.

Christian:
I want to ask a question giving an example problem.
Let us assume the mass of a distant Earth turns into light.
Does it require external energy, E= mass (of Earth) X (3x10^8)^2?

Einstein:
Yes.

Christian:
What would be the speed of that light on our earth?

Einstein:
$c = 3 \times 10^8$ m/s.

Christian:
I have a follow-up question.
Where did the rest of the external energy, the light speed squared, go?

Einstein:
A simple answer is—I do not know.
I can take a guess.
It must have been lost on its way to Earth.

Christian:
I wonder... how can empty space eat energy or light?

Einstein:
I do not know.

Christian:
May I ask where the leftover energy or light ends up?

Einstein:
In the black hole I saw.

Christian:
I have a final question.
Where is the black hole you saw?

The chair and editor:
The black hole is at the center of our galaxy.
We all know that—we published tons of articles.
In fact, the black hole discovery received a Nobel Prize.

That confirms the reality of a black hole.
And the black hole negates any invisible hand of 'god.'

Scene 7:
The great bomb maker Robert Oppenheimer (enters):
I am Oppenheimer…you all know me.
I am supposed to be a Marxist Communist.
I suffer from depression, my girlfriend died of suicide.
Yet my wife lives on.
I am the maker of killer bombs at Los Alamos.
The bombs I made work—ask the US Army.
The army used my bomb to kill tens of thousands in Japan.
We won the war to end all wars and created peace.
If I had not made the bombs first, the Germans would have.
I made the bombs when I lost my senses due to mental illness.
I studied the Bhagavad Gita…
And know I have become death as the destroyer of the world
kālo 'smi lokakṣayakṛtpravṛddho lokānsamāhartumiha pravṛttaḥ.
I am out of that business and took up the cause of "no bombs."

The chair and editor:
If there are no more Qs, this session concludes.
Oh, wait a minute.
I am told there are more experts waiting in line to be heard.
Let us hear them out too.

Scene 8:
The great Dawkins (enters):
I am Dawkins, an expert on "no god."
I became an expert studying biological sciences.
I am sure god does not exist, or there is no god.
Darwin and his principle of evolution is my "god."
Evolution controls all life—survival of the fittest works.

Please buy my book and read it to know more about my god.

Christian:
May I ask what biological sciences tell you there is no god?

Dawkins:
I study tissue and cells—you can too.
Tissue is a bunch of cells with atoms that do all the work.
There is no need for "god" anywhere.

Christian:
Who is making the "bunch of cells with atoms" work?

Dawkins:
I never saw a need to know that.
The "who" or "why" is for the confused birds!
I am not confused, and my "god" of evolution makes atoms work.

The great Oppenheimer (enters):
I am Oppenheimer, an expert in the study of genes.
I study African ape bones as a hobby.
I know African fishermen migrated into Israel.
I know their civilization settled in the caves and deserts.
It made them evolve as the "survivors of the fittest."
They devised the right-to-left script living in the caves.
They compiled the Torah and memorized it.
Why am I here? I want to tell you the real story.
A few Israel fishermen, having learned the Torah, went into India.
Believe me; I studied their bones found along the coast.
Evidently they taught their culture to the natives, who became
Hindu.
Hindu worshipped the fisherman leader, a Pharisee Krishna.
Later, the Hindu migrated to other parts of the earth.

The Hindu made the blue Krishna to appear as a white Jesus.
All this can be clear proof that man is from Africa.
The evidence that man evolved from an African ape has vanished.
The only evidence available in Africa is a bone left here or there.
I found one such bone and here it is—it proves the point.
I, an expert, made an induction based on this bone…
After all that Darwin was correct.
That makes Darwin the great ape-man god.
There is no place on the earth for other god, whoever that might be.

Christian:
I have a question.
Did you see the Israel fishermen walk the elephants and tigers into
India? Or did they take elephants and tigers in Noah's Ark later?
Do you see any left-over elephants or tigers in Israel's caves and
deserts?

Fig.4.4: The expert-Oppenheimer with a bone from Africa

Oppenheimer:

That is a good question, and I never thought of it.

I must think to come up with a good answer to that.

I know I can.

When I do, I will put it in my new book.

Keep buying my books—I need money.

The expert professor Schmidt (comes in):

I am Schmidt, an expert on the accelerating universe.

The universe is running away from us.

It is not an expansion of the universe that Hubble saw.

We know Hubble made Einstein place a magic constant in GR.

I saw the universe accelerating at the speed of light.

Two of my friends saw the same thing—that makes it a sure thing.

All galaxies are running away from us at the speed of light.

A voice:

It is great to see accelerating universe. Here is the Nobel Prize.

Schmidt:

Thanks for the Nobel Prize.

In a few billion years from now, our galaxy will be alone.

I will be here to verify it.

The astronomers on Earth will find no work to do.

You can trust my word as a "word" in the Bible.

I must point out that an old engineer tried to show that I was wrong.

That time I spoke to a large crowd of scientists in Washington.

I also spoke on this topic recently in Beijing.

I did not find any other nonbelievers.

Christian:

I have a question.

Do you know why the galaxies are running away from us?

Are they afraid of man because you saw them?
Are they trying to hide because you may laugh at their shapes or colors?

Schmidt:
I do not know. I should have asked them, but I forgot to ask.
I think dark energy might be a reason...
But no one has seen dark energy yet.
Einstein put forth a magic constant to describe curvature of space-time.
That magic constant might show us the dark energy one day.

Professor Mitrovica (comes in):
I am an expert on the earth's polar wander.
I went into the rocky earth to study and model true polar wander.
Polar wander is—a relative movement between surface and axis.
Earth's surface tips over and then returns to its original location.
Do I know why it does that—no, I have no clue.
I sat at the pole and saw the pole shift up to 50 degrees,
And it turned around—I saw the pole return close to its original location.
That is the process of true oscillatory polar wander.
I saw rock cooling in a magnetic field,
And saw the recorded properties of the field.
I decoded that in the lab and studied the magnetism in the rocks.
The properties of the field were recorded millions of years ago.
I measured changes in the orientation of Earth's magnetic field stored-
In ancient rocks and saw the effects of oscillatory polar wander.
I made a computer model with a combination of two mechanisms-
I saw in the model the wandering pole move back to its original location.
The first mechanism relates to Earth's equatorial bulge—

That equatorial bulge has a stabilizing effect.
The second mechanism relates to the strength of the tectonic plates.
When Earth's surface tips over relative to the rotational axis,
The twelve tectonic plates deform—plates want to go back to
original size.
These stabilizing elastic stresses play a role in the return of the
pole—
I saw that, it was the combination of both effects that did it.
I am surprised that it took about 10 million years to the pole.

A voice:
It is good to see. Surely deserves a Nobel—soon.

Mohammad:
Did you sit at the pole for 10 million years to find that out?

Mitrovica:
No, I did not.

Esther:
Does the equatorial bulge or the stress in the plates move counter
to polar wander? What about the fluid motions within the earth that
causes the wobble?

Mitrovica:
I do not understand the question.
I am sure our model is correct.
It makes no difference whether or not the bulge or stress can move,
or the fluid within the earth causes the wobble.

Esther:
What I meant is the bulge or the stress could not have moved
counter to polar wander and force the pole to return to its original
location.

Mitrovica:
I have nothing further to add.

Billionaire Yuliner (comes in):
I am Yuliner, a failed physicist…
But, I am a successful Wall Street gambler and made tons of money.
I do not know what to do with all this money.
I give away some of it to the scientists to do more fiction.
Man needs fiction to forget god…
And I need publicity to make more money.

Esther:
I am confused.
Why do you need more money if you do not know what to do with it?

Yuliner:
You think I am crazy? No, I am not.
Listen to me carefully. I am an addicted, greedy gambler.
The more I get small-time gamblers to invest, the more I feel good.
That is the only game I know to make money, and it works well.
Giving away a few peanuts to scientists means I get free publicity.
That makes the small-time gamblers come running to me and invest.
That gives me a chance to go for the kill and amass all the wealth.

Tobacco Smoky (comes in):
I am the Smoky, and make the smoke that is addictive.
And those who smoke kill themselves.
That does not bother me. I get rich.
All are destined to die anyway, so why think?

Killer Gunny (comes in):
I am the Gunny, and make killer guns that can blow up brains.
It does not matter if little people blow each other up.

That makes me rich and powerful.
All are destined to blow up one day, so why bother?

The Dalai Lama (enters):
I am Dalai, a poor farmer by birth.
But God has a plan.
God chose me as a servant to the monks.
The Marxist reds threw me out of my holy birth place.
I have become a roving ambassador for God.
I live in India, our holy motherland.
Buddha did not say much about God, but I believe in the Creator.
I believe in god-*Indra*, a reality.
I wrote a book based on the teachings of Nagarjuna of India.
Nagarjuna, a devout disciple of *Buddha*, taught in the East.
Jesus taught the same message in the West.

The chair and editor:
I assume there are no other experts.
Therefore this session concludes.
Kids, have a good summer holiday, and careful in your travels.

Fig. 4.5: Einstein, Darwin, Lamaitre, Hubble… sing and dance

Who knows the Universe, we do
Who knows the big Bang, we do
Who knows the expanding Universe, we do
Who knows curvature of space-time, we do
Who knows the black Hole, we do
Who knows black Hole eats all and spits Nobel, we do
Who knows making killer bombs, we do
Who knows no publish against big Bang, we do
Who knows no publish against curvature of space-time, we do
Who knows no publish against black Hole, we do
Who knows no publish against expanding Universe, we do
Who knows Man came out of an African Ape, we do
Who knows Ape bone is the best evidence, we do
Who knows no-god in the big Bang, we do
Who knows no-god in the black Hole, we do
Who knows god is not here or there, we do
Who knows no one sees a sign of god, we do
Who knows we lost our minds, no one, yet!

Chapter: 5

aatmaanam rathinam viddhi
sareeram rathamevatu
buddhim tu saaradhim viddhi
manaha pragrahamevacha

Know the Self as the Lord of the Chariot and the body as the Chariot itself
Know the intellect as the charioteer and the mind as the reins
(*Kathopanishad*)

Veda, the knowledge of Nature

That year in the summer Christian, Esther, and Mohammad arrived in India. They came for a visit to see and learn about ancient history. At the airport, they see a flashing broadcast. They pick up a local English newspaper with the headline "Dear *Rishi* saw *Isvara.*"

Christian wonders, "Who is this *Isvara*?"

Mohammad jokingly says, "Not me. No one saw me here just yet."

Esther says, "Let us find out what *Isvara* is."

Mohammad thinks, it could be a tropical tree or a drink that beats this heat.

Christian approaches a man and inquires, "Sir, can you tell me what this 'dear *rishi*' or '*Isvara*' means?"

The man responds, "Oh, that. Dear *rishi* is an incarnation of a deer, and *Isvara* is god, the Creator."

Surprised to hear that, Esther exclaims, "You mean a deer saw the Creator?"

Fig.5.1: Christian, Esther, and Mohammad in India at the airport

The woman standing next to the man calmly confirms the news. "Yes, here in India some animals come with more senses to listen, learn, and see *Bhagavan* (god) *Isvar*. I am not surprised if *rishi* saw *Isvara*."

Mohammad cannot believe what he has heard, and to make sure he heard it right he repeats it. "Really, you think *rishi* saw the Creator."

The man is sure. "Go ahead and ask the *rishi*. He will tell you."

Christian asks, "Where can we find this *rishi*?"

The man gives them directions to the *ashram*.

Mohammad does not believe this story. Therefore he is anxious to go on with their planned trip and not take a detour. "Hey, guys, wait a minute, we did not come here to see a *rishi*, a magician."

To make Mohammad relax a bit, Esther cracks a joke. "You mean a magician like Moses."

Mohammad is still concerned about taking a detour and wasting time. "Yes, they are all magicians. Ask your rabbi, the chosen one. He will tell you all about it. Do we know why Moses did not name the 'god' of the Ten Commandments?"

Esther says, "I think the reason was that Moses wanted to keep it as a secret so the 'god' could be 'whoever that might be.' That was then, and now we are in the present time, and I am curious to find out if this *rishi* is a magician."

Mohammad gives up. "OK, majority rules. Let us get it over with quickly. We have to visit many places in a short time in this hot climate."

They take a train, then a bus, and finally walk to reach the *ashram*.

They leave their shoes at the entrance, go inside the *ashram*, and sit on the floor with others. The *ashram* is full of people who sit on the floor singing songs.

Om Brahman, Om Isvar
Om Brahma, Om Narada, Om Vishnu-Laksmi, Om Shiva-Parvati
Om Indra, Om Surya, Om Bhoomi,
Om Sita-Rama, Om Radha-Krishna
Hari Rama, Hari Krishna…

At the end of the singing, all stand up and one by one walk up to the *rishi* to receive blessings and *prasadam*. Christian, Esther, and Mohammad also participate in this ritual. The *rishi* acknowledges the three visitors and welcomes them as guests.

The *rishi* requests the guests to take seats and makes an announcement. "Today we have three young visitors. Let us help them with their needs." The *rishi* looks at the visitors and says, "Please feel free to ask any question."

Mohammad did not understand any of the words in the songs. Therefore he begins the inquiry with, "What is *Om*?"

The *rishi* defines *Om* as the *shakti* of the vibrating sound that creates or destroys.

Esther wants to know, "What is its origin?"

The *rishi* tells them that the origin of the vibrating sound is *Isvar*.

Christian asks, "Where is *Isvar*?"

The *rishi* answers that question in a broader sense, saying, "*Isvar* is present in all places. All you need is to pick a place and look to find *Isvar*."

Esther points to the eternal flame. "Is *Isvar* here?"

The *rishi* affirms, "Yes."

Mohammad pursues the issue further. "I do not see *Isvar* in this plant or there in that idol."

The *rishi* makes it clear that this means he is not ready yet. "And by the way, it is a *murti*, not an idol."

"What does it take to be ready?" Esther asks.

The *rishi* suggests that they should study the Vedas and learn about the *param-brahman*.

Mohammad starts to wonder and asks, "Where is the *param-brahman*?"

The *rishi* looks up. "We live within the *param-brahman*."

Christian has a follow-up question: "How do I know that?"

The *rishi* suggests that Christian should spend time to understand the message of *Krishna* in the *Bhagavad Gita*.

Mohammad asks, "Was *Krishna* a Pharisee like Jesus?"

The *rishi* responds, "No, *Krishna* is not a human like Jesus. *mah-Isvar* in the form of *Krishna* spoke to us in the *Bhagavad Gita* for the *param-brahman*."

Mohammad wants to pursue a new topic. "If you can see *Isvar* and know we live within the *param-brahman*, why do you live in this hut?"

The *rishi* poses a question. "Where should I live?"

Esther says to the *rishi*, "Why don't you come to America, sell knowledge, and get rich?"

The *rishi* has another question. "What is rich?"

Mohammad knows it all, or so he thinks. "Rich means money and power."

The *rishi* raises a philosophical question. "What use is money or power?"

Esther thinks she knows the answer to that. "It gives security and happiness."

The *rishi* wants to know what they are thinking. "Am I insecure and unhappy here?"

Mohammad says, "I think with money you can be more secure and happier."

The *rishi* informs the young visitors, "Always do your righteous duty and do not expect success or failure. That alone makes you secure and happy, not the money or power. Buddha, Jesus, and the rest that came out of a hut were happy within the self. Those who learn to do the righteous duty to the *param-brahman* with the knowledge of nature, the *param-brahman*, or *bhakti* (prayer) toward the *param-brahman* feel secure and happy. The knowledge of nature is for the self, not for selfish use or to get rich or become powerful and control other living beings."

Mohammad inquires, "Then what is in store for less than one percent elite that own the military-industrial complex, the banks, buildings, land, crops, radio and TV, and control the trade of goods fixing markets, and gamble on Wall Street or get rich and powerful with greed, and control the rest of more than ninety nine percent of living beings? Are those righteous acts?"

The *rishi* clarifies: "One can own or manage to the extent of one's ability for the benefit of all, but the ability given by *Isvar* is for the self, not for selfish purpose or to get rich and satisfy one's own senses."

Esther raises another issue. "What about the one that forms cults in the name of god or claims there is no god or can only see god in money and power? What about those that lie, cheat, create chaos, and kill for the thrill of a game?"

The *rishi* makes it clear that such acts are unrighteous and add bad *karma* to the self. "And the self with bad *karma* faces consequences in this or the next birth," he explains. "The self of the victim that suffers injury accumulates good *karma*."

"Is the self, consciousness, *atma*, and *Isvara* the same?" Esther asks.

The *rishi* nods. "Yes."

Christian wonders why a living being would do unrighteous acts.

The *rishi* says, "Prior bad *karma* controls the senses of that living being in the present birth. That way the consciousness becomes out of tune and fails to receive signals properly from the consciousness of the *param-brahman*, like an out of tune radio or television that will not receive broadcast signals properly from a station. That is true of a selfish ruler, king or queen, judge, or even a worker. The self of an *adharmic* always faces bad *karma* and has to go through many birth-death cycles and suffer before becoming purified. The ninety nine percent and the one percent will have the same chance to do *dharma* learning the knowledge of the *param-brahman* or *yoga* or *bhakti*, for the purpose of the self and attain *moksha*."

Christian summarizes the concept to assure himself that he understood it. "That means an act for the self is an act for the purpose of *Isvara*." And he raises a related issue. "What about those that follow a religious faith?"

The *rishi* explains, "If one allows the senses to pursue a desire for selfish reason, it results in bad *karma*; if one pursues a desire for the purpose of self, it results in good *karma*. It does not matter whether success is evident to the senses, because the self keeps the record. Therefore, how one uses the senses or free will matters. The ability to control the senses comes from birth, and it relates to one's prior *karma*. To control the senses from pulling in all directions, one must learn to meditate or concentrate the mind on an object, a process known as *yoga*. The object can be any creature, man, or woman, or the *murti* of *Isvar*. Mastering such a practice would enable one to do one's righteous duty. Religion is a creature of an insecure *manu* who thinks 'god' sits at some place and makes it all happen. Such a faith would not help control the senses or enable one to do his or her righteous duty."

Mohammad comes up with another issue. "If a man with ability cannot enjoy the fruit, why should he work hard to create the fruit that he cannot enjoy?"

The *rishi* answers this way: "The issue is whether you are here to keep satisfying the desires of your senses or to do your righteous duty. Your ability comes from *Isvar*, and that ability is to purify the self for the use of the *param-brahman* who created you, like the house you built for your purpose."

Esther reminds them that the self or *Isvara* is within the living being. "Why would the self and the senses pursue different paths otherwise?"

The *rishi* explains the inherent conflict this way: "It is a very good question. There is a constant struggle involving divided loyalties, like a battle between good and evil, or god and the devil. This is what is happening within your body. Self belongs to the *param-brahman*; desire of the sense(s) is the devil that belongs to your body. The goal of the self is to get purified and leave the body, whereas the goal of the desire is to keep the body well and preserve it as long as possible and house the self. Clearly there is a conflict and a struggle here. 'Free will' allows the senses to go in all directions and obtain whatever is needed to benefit the body and enable it to house the self for eternity. The self knows this will not work and prefers *moksha* rather than staying in the body as a prisoner.

Therefore one has to learn to steer the sense-horses that pull the chariot or the body in a proper path as *Krishna* tells us in the *maha-Bharat* War and win to allow the self to purify. Then *Isvara*, as the 'pure self,' can go on to serve the *param-brahman* directly."

Mohammad wants to know whether the caste system here and the class system in a beehive or ant colony is selfish.

The *rishi* clarifies the issue this way: "It is the *verna*, not the caste. *Verna* defines the ability to do a duty, and one's *verna* duty can benefit all, including the self. A duty performed under *verna* is righteous, like the duty performed by a worker in a corporate system."

Christian asks, "What is hell and what is heaven?"

The *rishi* explains it this way: "The 'cloud' density, within the self, increases when the self accumulates bad *karma*. That self that can't be purified here on the *bhoomi* will be reborn in another *graha* of a different universe where conditions will be much more severe. Either that or the self will be sent to *Surya Deva* to burn that excessive 'cloud.' That process appears as if it is the *naraka*, or hell.

The 'cloud' density, within the self, decreases when the self accumulates good *karma*. In this case, the self gets purified here on the *bhoomi*, up to a limit; then it will undergo rebirth in another *graha* of a different universe where conditions will be much better to achieve further purity, or if purified on the *bhoomi*, it goes on to join *mah-Isvar* in the service of the *param-brahman*. It is like your purified blood going into your brain to serve you. That process appears as if it is the *svarga*, or heaven."

Christian summarizes his understanding with a question. "It looks like the process of purifying the self involves several steps, like my blood going into the lungs, liver, kidneys, etc., before it gets purified and becomes fit to serve in my brain and other organs. Am I correct?"

The *rishi* affirms Christian's summary with a nod.

Mohammad says, "That means the unrighteous will end up in hell at some stage. Am I correct?"

The *rishi* clarifies, "Yes, if one keeps on accumulating bad *karma*."

Esther brings up some famous scientists with a question. "Darwin is for survival of the fittest. Marx is for class struggle and believed 'false consciousness of the elite sees social order as predetermined.' Is that true?"

The *rishi* reminds them of the role played by the senses in this struggle for survival of the fittest. "Yes, the ability to do one's duty and survive is predetermined by *Isvar*; and that ability defines one's *verna*."

Christian goes back to an earlier point made that Buddha, Jesus, and the rest "came out of a hut." "What is the significance of that?" he asks.

"They learned the Vedas in a hut or under a tree," the *rishi* answers, "and performed their righteous duty to help purify the self."

Mohammad raises a question. "If Jesus was a Pharisee himself, why did Jesus and the Pharisees differ?"

The *rishi* explains it this way: "It shows that ability differs between any two individuals. Here, the ability of the Pharisee allows the senses to run away on the path of free will working for selfish purpose, whereas the ability of Jesus allowed the senses to follow a proper path of believing in the Father and performing his righteous duty to help purify his self."

"Who is the Father that Jesus referred to?" Christian asks.

"*Surya Deva* is our Father, unless Jesus meant the *param-brahman*," the *rishi* informs him.

The session ends, and the people leave. Christian, Esther, and Mohammad decide to stay and continue the process of learning. After a short break, the *rishi* resumes the class.

Christian raises a doubt, asking, "How do I know we live within the *param-brahman*?"

The *rishi* simply states, "We live within an *Anuvu* that belongs to the *param-brahman*."

Esther follows up with her own doubt. "How do I know I live within an Atom?"

The *rishi* starts to explain, "Our *bhoomi* orbits around the *surya*."

Esther agrees. "Yes, I heard Kepler and Newton say that."

The *rishi* continues, "And the *surya* orbits too."

Esther repeats what she heard. "Yes, the sun orbits around a black hole, which is at the center of our galaxy."

The *rishi* continues explaining, "The *surya* orbits around *Indra*, and *Indra-surya-bhoomi* plus seven other *grahas* together make one *Anuvu*."

Fig.5.2: Map of Earth with S-pole at 90E (after pole-reversal), *Rishi* explains map to Christian, Esther, and Mohammad

The *rishi* walks closer to the maps on the wall, and Christian, Esther, and Mohammad follow. The *rishi* points to one map at a time, explaining how the spin of the *bhoomi* relates to a *yuga* of 4,320,000 years and the pole-reversal.

The *rishi* begins with a map that shows the South Pole of the earth pointing toward the sun. (See Fig. 5.2). "At that time, life on *Bhoomi* would be sparse or nonexistent. The hemisphere that faces the *surya* gets light hot days with no light at the equator and beyond or at the polar region of the other hemisphere. Therefore the other hemisphere mostly gets dark ice nights with ice accumulation. Such a condition continues to exist for thousands of years. During that time, life may exist in certain spots, conducive for that life, between the polar region of the hemisphere that faces the *surya* and the equator region.

At that time, *Bhoomi* rotates on its axis only once in each orbit around the *surya*. Actually, a similar condition is due to occur in about 700,000 years from now as the present *yuga* comes to an end as pole-reversal nears."

Fig.5.3: Map of Earth with S-pole at 135E, *Rishi* explains map to Christian, Esther, and Mohammad

The *rishi* continues to explain nature using the next map. "*Indra* makes the North Pole of *Bhoomi* flip to the South Pole, a pole-reversal as shown in Fig. 5.2. *Indra* sets a new axis of rotation for *Bhoomi* as the North Pole changes into the South Pole. That makes the *bhoomi* start rotating slowly with a wobbly spin. Thus it is the beginning of a new *yuga* with the South Pole spinning clockwise, or away from *surya*. In this new *yuga*, the phase one period will be 1,728,000 years. In this phase, the South Pole spins 45 degrees clockwise or away from *surya*, from 90E to 135E. (See Fig. 5.3). The North Pole of *Bhoomi* spins 45 degrees clockwise, or toward *surya*, from 270W to 315W. In this phase, during the first part of the 22.5-degree spin, new life begins on *Bhoomi*, and that life slowly multiplies in certain areas conducive to life, like the regions between the equator and halfway toward the poles. During the second part of the 22.5-degree spin, the life multiplies and migrates into other parts on the *bhoomi*."

Fig.5.4: Map of Earth with S-pole at 180S, *Rishi* explains map to Christian, Esther, and Mohammad

The *rishi* then goes to the next map with the South Pole at 180S, or the North Pole at 0N (360). "In this new *yuga*, the period for phase two will be 1,296,000 years. In this phase, the South Pole spins 45 degrees clockwise, or away from *surya*, from 135E to 180S. (See Fig. 5.4). The North Pole spins 45 degrees clockwise, or toward the *surya*, from 315W to 0N (360). In this phase, during the first part of the 22.5-degree spin, life continues to multiply on the *bhoomi* in the areas between the equator and closer to the poles. During the second part of the 22.5-degree spin an ice age sets in at the polar-regions. That makes the conditions for the survival of life diminish on *Bhoomi* between the poles and halfway to the equator due to the accumulation of ice peaks in the polar-regions. That condition forces the life in the polar regions to migrate back into the equatorial regions for survival, where a hot age sets in."

Fig.5.5: Map of Earth with S-pole at 225W, *Rishi* explains map to Christian, Esther, and Mohammad

The *rishi* goes to the next map that shows the North Pole at 45E, or the South Pole at 225W. (See Fig. 5.5). "In this *yuga*, the present phase three period is 864,000 years. In this phase, the South Pole spins 45 degrees clockwise, or away from the *surya*, from 180S to 225W. The

North Pole spins 45 degrees clockwise, or toward the *surya*, from 0N to 45E. In this phase, during the first part of the 22.5-degree spin, completed about 30,000 years ago, the ice accumulation continued in the polar-regions and the hot age continued (and will continue) in the equatorial region. That condition allowed life to continue multiplying in the equatorial region, such as in *Bharat*, during the past 30,000 years and migrate into the east, west, south, and north as the ice began to melt. During the second part of the 22.5-degree spin—or the 1-degree plus 22.5-degrees spin equals to 23.5-degrees spin at this time—the polar ice is melting away with the spread of the hot age into the polar regions. Thus the conditions for survival of life on the *bhoomi* will begin to diminish during the next 250,000 years."

Fig.5.6: Map of Earth with S-pole at 270W, *Rishi* explains map to Christian, Esther, and Mohammad

The *rishi* explains the last map hanging on the wall, which shows the North Pole at 90E, or the South Pole at 270W. (See Fig. 5.6). "In this *yuga*, the final phase four period will last 432,000 years. In phase four, the South Pole spins 45 degrees clockwise, or away from the *surya*, from 225W to 270W. The North Pole spins 45 degrees clockwise, or toward the *surya*, from 45E to 90E and faces the *surya*. In phase four, during the first part of the 22.5-degree spin, the ice melts away and the sea level rises with all coastal areas on the *bhoomi* going underwater. The hot age moves into the polar-regions and makes life on the *bhoomi* sparse. During the second part of the 22.5-degree spin comes the end of time, in about 700,000 years from now. That results in a light hot age at the North Pole region with no visible light at the equator and beyond, and a dark ice age sets in at the South Pole region. That makes life on the *bhoomi* sparse or nonexistent for thousands of years until *Indra* flips the North Pole into the South Pole, a pole-reversal, and the new axis of rotation of the *bhoomi* begins to wobble as it spins."

The *rishi* explains about an *Anuvu* by looking into the Veda.

Christian comments on the fact that Newton missed the role of *Indra* and saw the sun as the lone player in rotating and orbiting the earth. "Thus he failed to see the spin of the earth."

Mohammad agrees and points out, "Einstein missed it too."

Esther says, "I think no one else saw it."

The *rishi* informs them that the ancient *rishi* saw it before the last ice age.

Fig.5.7: *Rishi* teaches a part of Veda to Christian, Esther, and Mohammad

Christian gives a summary of the time period for the last ice age: "The North Pole of Earth spinning 45 degrees from 337.5W to 22.5E defines the ice age period. Am I correct?"

The *rishi* agrees. "Yes, the peak of the ice age lasted for over ten lakh (1,000,000) years. And prior to the ice age, or more than twenty-five lakh (2,500,000) years ago, *manu* began to migrate and spread into other parts on the *bhoomi* from the equatorial region of *Bharat*."

Christian says, "In other words, we had more than one million years of ice age, and the migration of man must have begun about two and a half million years ago. So what must have happened during the ice age?"

The *rishi* gives a summary of what happened. "During the last ice age we lost the ancient Vedas and related writings. We lost life in the polar- regions due to extreme ice and the cold climate. We lost life in the equatorial region due to extreme heat and a hot climate. Those that migrated away from the polar region and survived the ice age emerged with lighter skin color, and those in the hot climate emerged with darker skin color."

Mohammad says, "I see a few lighter skin-colored persons even in India. Where did they come from?"

The *rishi* says, "They are the siblings of those that came back to India during the ice age or later. *Manu* from *Bharat* migrated into the Northern and Southern Hemispheres before the last ice age. Some of their siblings came back into *Bharat* during and after the ice age."

Esther asks, "Did they become Hindus or Brahmins?"

The *rishi* says, "Some of them practice the Hindu way of living. By the way, it is the *sanatana dharma*. The name Hindu was derived from 'Indus' or 'Sind' river by Alexander the Great about 2,400 years ago."

Christian gives a summary of his understanding of the information from the maps. "Life multiplies and spreads on the earth in phase one of the *yuga* during the first 45 degrees of the spin; then comes phase two, the second 45-degree spin, with the ice age that destroys most life, leaving some life in the equatorial region. In phase three, during the third 45-degree spin, life multiplies and spreads; then comes phase four of the *yuga*, the final 45-degree spin with the hot age and the end of time, which destroys most or all life. Am I correct?"

The *rishi* says, "That is correct. You understood it well."

Christian follows up with a question. "The polar regions had no heat or light for more than one million years during the ice age, or heat and light could not migrate into the polar regions. What makes heat or light travel in space?"

The *rishi* explains, "The *paramanuvu* in *akash* that acts as a vehicle and allows heat or light to move. All other signals of the *shakti*, such as consciousness signals, the cold/heat force signals, nerve/electronic or radio/TV signals, light/vision/sound signals, and the smell/taste signals, move within *paramanuvu*."

Esther wants to know about this *paramanuvu*.

The *rishi* explains, "The *paramanuvu* is an infinitesimal *anuvu* that comes in infinite sizes. It fills the *akash*, the space in an *Anuvu* between *Indra-Surya-Bhoomi-grahas*. Also, it fills the space in an *anuvu* between neutron-proton-electron."

Christian brings up a point: "In science class, I learned that man can see or sense only about 4 percent of the matter in our universe, and about 96 percent of the remaining matter is '*dark*,' or man cannot see or sense it. Do you think that this 96 percent is the *paramanuvu*?"

The *rishi* goes into more details. "*Paramanuvu* is the matter that *manu* can't see or sense. It is present within you, me, the tree, and the *bhoomi*, *surya*, and *Indra*. The structure of *anuvu* and *paramanuvu* should be identical. The space in an *anuvu* is filled with yet smaller *paramanuvu* of infinite sizes. *Isvara* assembles a living being using a variety of *anuvu* and *paramanuvu*."

"How can I be certain it is true?" Christian asks doubtfully.

The *rishi* explains. "On the *bhoomi* alone we have eighty-four lakh (8,400,000) types of species, on land, in the water, and in space. Each type comes as male and female in large numbers, and each member of the specie comes with its own unique features. In other words, no two individuals look alike in all respects. And each individual comes with many types of organs, such as the skin, hair, nails, bones, muscles, nerves, brain, eyes, ears, nose, heart, blood, lungs, liver, kidneys, and stomach, etc. Each organ consists of a large number of *anuvu* and *paramanuvu*. That should make it clear there must be an infinite number of *anuvu* and/or *paramanuvu* that *atman* (*Isvara* or self) within assembles into a living individual to perform a specific duty."

Mohammad raises a question. "How does this fit into an ape evolving as a man per Darwin?"

"The body features do not evolve directly from an ape into a man," the *rishi* informs him. "The *anuvu* and *paramanuvu* from a dead ape could go into making a Darwin. When a fisherman eats a dead fish, some fish *anuvu* could become part of the fisherman, but the fisherman would not become a fish by eating fish all the time."

Esther says, "Scientists like Hoyle and Crick thought life may have had an extraterrestrial origin. Is it true?"

The *rishi* nods. "*Atma* within a living being has extraterrestrial origin. It assembles the living being using the *paramanuvu* made in *akash*, and *Indra-Surya-Bhoomi* nourishes that living being to do a duty."

Christian asks, "If *Isvar* creates all living beings, when or how did I become a living individual?"

The *rishi* explains it this way: "The *jiva* or *atman* within your body is eternal being a small part of *Isvar*. It exists in your present body because *Isvar* created the 'egg cells' in your mother's organ giving half-*shakti*, and created the 'sperm cells' in your father's organ giving another half-*shakti*; and by joining the sperm with the egg *Isvar* made that into one *Shakti* of *Isvara* or *atman* to begin the process of purification. The moment the sperm joined with the egg, a living individual body, you, began to form."

Christian wonders about the purpose of all this creation.

"The purpose," says the *rishi*, "is to get the *atman*, the conscience, purified for the purpose to re-unite with pure-*Isvar* and serve the *param-brahman*."

Christian wonders, "Does this mean that the *atman* with which I was born is not pure?"

"Not pure enough to be useful to the *param-brahman*. You are a little creature living within the *param-brahman*, like the little creatures that live within your lungs and liver that purifies your blood for your purpose. In the same way, you purify that *atman*, a small part of *Isvar* which is within you, for the purpose of the *param-brahman*."

Mohammad comes up with an analysis. "It is like the good bacteria or microbes living within to get me going."

"In fact, both good and bad bacteria live within," Esther agrees.

Mohammad makes a joke. "You mean like the gentile and the Pharisee."

The *rishi* points out the reality of nature. "Some living creatures within you work to keep your body performing, and some creatures within you work to destroy your body. It is a constant struggle until your time is up."

Christian wonders if one can look at the living beings on the *bhoomi* as good and bad bacteria of the *param-brahman*.

"Yes," the *rishi* agrees. "You can see how a few selfish can destroy the quality of air, water, land, and nature. *Krishna* in the *Bhagavad Gita* spoke of the righteous duty that every living being within the *param-brahman* must do. That message is equally applicable to the living creatures within you. The struggle between righteous and unrighteous continues until the end. Those that work to keep your body performing until you attain purity of the *atman* fight against those that work for your quick demise. And you will never know when that struggle may end, the struggle between those that want your body to keep performing and those that want it to end."

Christian asks for clarity as to whether *Krishna* as the consciousness of the *param-brahman* spoke about the struggle within the *param-brahman* as well as within the living beings on the earth.

"Yes," the *rishi* affirms. "*Krishna* as *mah-Isvar* within the *param-brahman* spoke about what righteous duty is and why one should act without expecting a reward for doing so. And *Isvara* within you gives a similar message to the creatures that live and work within you."

Christian goes to a new topic. "How am I able to sense light, sound, smell, taste, and touch?"

The *rishi* explains, "Each sense organ in your body is made of certain types of *anuvu* and *paramanuvu*. A particular type of *anuvu* in your eye acts as the receptor for light. Similarly, a certain type of *anuvu* in your ear acts for the sound; those in your nose act for the smell; those in your tongue act for the taste; those in your skin act for touch."

Christian repeats the summary of the concept. "So the *paramanuvu* in an *anuvu* in my eye acts as a receptor and accepts the incoming light, and the nerve in my eye transmits that light signal into my brain. Am I correct?"

The *rishi* agrees. "Yes, the nerve that transmits is *narada.*"

Christian asks, "What makes me sense the light or see the object?"

"It is the *Isvara*, the consciousness within, that enables you to sense the light or see the object," the *rishi* says.

Mohammad is not clear what the consciousness is.

The *rishi* explains, "*Brahma* is the consciousness of the *parambrahman*. A small part of that consciousness in the form of a *paramanuvu* cloud is the *jiva* (*atma* or self or soul) that resides within you. The density of this 'cloud' varies between any two individuals in the same specie or between any two species, and the density in each individual sets that individual's ability to receive signals, think, and act. It also sets the shape, size, and the looks of that individual. The self exists on the *bhoomi*, solely to get *chittashuddhi*, or get rid of that cloud for a purified self, a requisite for *moksha*."

Christian asks, "Is the pure self like the pure light with no cloud of mass? If so, one must do the righteous duty without fear of death and attain *moksha*, correct? And where can be this self in a living being?"

"Yes, the pure self is without cloud. The fear of death or the desire to live longer is due to insecurity and ignorance of the reality caused by the senses that direct free will. The self is within the brain and shines as the mind."

Esther points out what she has learned in a science class. "I learned there is an area called the thalamus-hypothalamus with a special center known as the supra-chiasmatic nucleus, the SCN, which initiates signals to other parts of the brain. The SCN is a pinhead-sized brain structure that contains about 20,000 cells and is the control center for all signals. Can we call that a god particle?"

"It is reasonable if one wants to call it a god particle," the *rishi* says. "*Isvara* in the form of the *paramanuvu* cloud could reside in only one of those 20,000 cells that could drive all other cells in the body. You cannot see or sense *Isvara* other than in the acts."

Esther thinks back to the song the disciples sang. "Can you tell us what *Brahma, Narada, Vishnu-Lakshmi,* and *Siva-Parvati* are?"

The *rishi* explains, "*Isvar*'s *shakti* is *Brahma*, as the consciousness; *Isvar*'s *shakti* is *Narada*, as the consciousness signal; *Isvar*'s cold *shakti* is *Vishnu*, as the creator of the *paramanuvu* union; and *Isvar*'s hot *shakti* is *Siva*, as the destroyer of the *paramanuvu* union. The live *anuvu* and the *paramanuvu* within act under the *shakti*. *Isvar* provides that *shakti* while *Brahma* with consciousness transmits signals through *Narada* to the cold and hot *shakti* to act. *Lakshmi* makes *Vishnu* a complete one-*shakti*; so is *Parvati* to *Siva*. It is like the sperm cell joining the egg cell to create one-*shakti* of *jiva*."

Christian says, "I learned in science class that the sperm comes with twenty-three chromosomes, and when it joins with the twenty-three chromosomes of the egg, the egg fertilizes and life begins. Is that the reason to give a male-female name as *Vishnu-Lakshmi*?"

The *rishi* agrees with a yes.

Esther inquires if there is any significance to the name *Sita-Rama* or *Radha-Krishna*.

The *rishi* explains the significance this way: "In *yuga* phase two, *Rama* represents the righteous nature of a male, and *Sita* represents the righteous nature of a female. In phase three, or in the present time, *Krishna* represents the nature of a male, and *Radha* represents the nature of a female. At present, we are in the last quarter of phase three in this *yuga*. In phase four, this *yuga* comes to an end of time."

Mohammad raises an important question on the color of the deity. "Why does the *murti* appear dark for *Vishnu, Siva* or *Rama, Krishna*?"

The *rishi* explains, "*Manu* cannot see or sense most of the *param-brahman* that appears dark. *Manu* cannot directly see or sense *Isvar* and *Indra* other than to observe the actions. *Manu* can sense cold and see dark from *Vishnu*; sense heat and see light from *Siva*. So, the dark *murti* represents *Isvar* and the rest that is invisible."

Christian wants to know how *Vishnu* relates to *Indra* and *Siva* relates to the *Surya*.

The *rishi* explains, "*Vishnu* as the cold *shakti* manifests the *paramanuvu* union within *Indra*. That means *Vishnu* joins with *Siva* to become part of *mah-Isvar*. *Siva* as the hot *shakti* manifests the destruction of the *paramanuvu* union within *Surya*. That means *mah-Isvar* gets divided into *Vishnu* and *Siva* with one-half of *Brahma* and one-half of *Narada* going each way. The *anuvu* made of the *paramanuvu* union moves from *Indra* toward *Surya*, and the destroyed union of *anuvu*, as the *paramanuvu*, moves from *Surya* toward *Indra*. It is like the cold air from the sea moving toward land

and the hot air from the land moving toward the sea. Now are you satisfied we live within an *Anuvu*?"

Christian says, "I am convinced, sir. I wish to see more on how *Indra* controls the sun, earth and planets etc. in an Atom."

The *rishi* informs them that the Pharisee-Persian took parts of the Veda. "One of the parts in it will show you the details of how *Indra* controls the *surya*, *bhoomi* and *graha*s. I understand a part of the Veda remains buried in an ancient temple in Kashmir."

Mohammad says, "I heard news that a professor tried to uncover Jesus's bones in Kashmir and wanted to dig at a masjid, an ancient temple, in Kashmir. Do you think the Pharisee may have buried the Veda there?"

The *rishi* makes a suggestion. "You have to go and look. If it has any link to the Pharisee-Persian then it needs some attention."

Mohammad says, "Kashmir is one place I always wanted to visit."

The *rishi* makes another suggestion. "Before you embark on that adventure, learn ancient history a bit more. That might help you get to the right place to look."

Esther asks for advice. "Please tell us where we can get a crash course on ancient history."

The *rishi* thinks of a place. "I can suggest a good place, an ancient temple not too far from here. You may go there and ask the *swami* for help. You will find pieces of ancient history in the records kept there, and if you search the old records patiently you can learn a lot."

Christian, Esther, and Mohammad take blessings from the *rishi* and move on with the adventure of learning and finding the missing part of the Veda.

Glossary:

Bhagavan=Almighty *Brahman*

Adharmic=Doer of injustice or unrighteous duty

Grahas=Planets

Akash=Space

Swami=Learned teacher

Deva=Nature of god

Devi=Nature of goddess

Sanatan dharma=Eternal law of nature

Lakh=100,000

Chapter: 6

yovaibhoomaa tat sukham nalpesukham asti

The infinite is happiness.
No happiness in anything that is finite
(*Chhaandogya Upanishad*)

The ancient-History of Man

Christian, Esther, and Mohammad meet the *swami* in an ancient temple. And with the *swami*'s help they search and review the old records, puppet show clips, and sketches. Thus they learn the gist of relevant ancient history as the *swami* reads the old text.

Christian finds an old sketch and asks the *swami* to see if it depicts the prince as *Buddha*.

The *swami* looks at it and nods. "Yes, this is a sketch that shows *Buddha* teaching his disciples."

Esther asks the *swami* to read the script below the sketch and translate it to know what message *Buddha* taught his disciples.

The *swami* reads the message written in Pali script, which is a child of the Brahmi script, and translates it into English, which also is a child of Brahmi, where *Buddha* said:

"All life is One, and *dharma* prevents *manu* from harming any living being, even a bee or ant. Every living creature has a purpose and performs a duty to the Creator. Discover the good, and use that good for the benefit of all. O disciples, go teach the message in the East and in the West."

Mohammad thinks the message is the same one that the Veda teaches.

The *swami* agrees. "Yes, that message is the guiding principle in the East, and it is equally applicable to all living beings on the *bhoomi*."

Fig.6.1: Prince as the enlightened-*Buddha* with disciples in Himalayas

Christian wants to keep searching to know whether that message came to the West.

Esther finds something and wants to know if it is helpful or not. The *swami* looks at it and informs her that it deals with an ancient history of an ancient Pharisee-Persian.

Esther says, "I think that sounds like my ancestors' history."

Mohammad says, "I think that history belongs to my ancestors too."

Christian says, "I have nothing in common there."

The *swami* assures Christian that this is not true. "Your ancestors too came from the same root."

Christian inquires where that root is. "Is that in Africa?"

The *swami* says, "That is what the scientists think. It is Africa because they see various types of people there as well as ape bones."

Christian asks, "Is that a good basis to conclude man comes out of Africa?"

The *swami* shakes his head. "No, because I see primarily two types of people in Africa looking at the facial features, the nose, lips, hair, etc. The third type is a mix of the other two. The facial features of the people in Africa's east and north belong to type one, the people in Africa's west, middle, and south belong to type two, and the rest in Africa are a mix of type one and two. The rest of the people on the *bhoomi*, nearly 90 percent of the total people, look closer to type one."

Esther wonders what all this meant.

The swami concludes, "That primarily there are only two types of people on the *bhoomi*. About 90 percent of the total people come from type one plus the mixed; and about 10 percent of the total belongs to type two plus the mixed. Since type two is primarily concentrated in Africa, it is the root for type two, and since type one is primarily concentrated in southern Asia or *Bharat* (*Dravida* land) it is the root for type one."

Esther locates another piece of the puzzle and gives it to the *swami*. "I found this. What does it say?"

The swami begins to read:

"The soldier of an ancient king became a Persian.
A disciple of an ancient rishi became a Pharisee.
The Persian won a battle and returned to the palace.
The Persian reported the good news to the Pharisee."

Esther jokes, "Oh, so my ancestors owned palaces."

Mohammad counters that with a joke, "Well, my ancestors won battles."

Christian is interested to find out more. "Let us see what comes next."

The *swami* translates the conversation between the Persian and the Pharisee that appears on the sketch. It goes like this:

Persian:
"Salam, high priest of asura, the god of senses, we did it.
All land from the Indus River to Turk to Egypt is ours.
No one can ever challenge us.
We will go on to defeat the sura, the god of righteousness.
We will take over the earth."

Fig.6.2: Moses with Plaque standing before his people

Pharisee:
"What did you do with Moses and my brothers in Egypt?
Did you let them go free?"

Persian:
"Yes, master.
Moses burned to death on the way to Israel.
The Ten Commandments did not help the magician.
The burning bush behind did it."

Pharisee:

"O Persian, you did it well.

I proclaim you as the mighty emperor of the Asura.

We must re-educate the gentile in our kingdom.

We must teach the gentile the 'right to left' writing.

I devised the new script from Brahmi using my free will.

We must teach the Avesta to the gentile.

We must destroy the Veda, the idols, and the Varna.

You must go on and defeat the brown gentile.

Remember, no harm comes to the rishi."

Fig.6.3: Persian and Pharisee in a palace

Persian:

"Yes, master.

I will, with your blessings."

The *swami* reads the song that the Pharisee and Persian sang to celebrate the win in battle, and the song goes like this:

We won we are the chosen
We won we have the free will
We won we are pure with lighter-skin
We won we are superior
We won we make the rules
We won we rule over all life
We won we write *Sanskrit* in Brahmi from right to left
We won we devise Avesta from Veda
We won we know of no-gods or Idols
We won we know god-*Asura*, whoever that might be
We won we pray god-*Asura* to make us rich and powerful.

The *swami* keeps on reading the script that describes the continued battle of the Persian. A sketch shows the chaos, and it reads as follows:

"The Persian sets fire and burns the ashram.
Chaos prevails all over the ancient kingdom with the destruction.
Apes, monkeys, and other animals run away looking for safety.
The Persian kills the cow, calf, goat, and deer, and takes for a meal.
The elephants and tigers on the other side of the river watch."

Christian locates another sketch and gives it to the *swami*. "What is this sketch? Is it another Persian battle?"

Fig.6.4: Persian battles at the *Ashram* and kills the cow

The *swami* looks at it. "Yes, it is a battle. It is between Alexander of Macedonia and the Persian."

Mohammad becomes curious. "What does it say below the sketch? Did the Persian start this battle too?"

Fig.6.5: Alexander defeats Persian and burns the Palace and Avesta

The *swami* reads the script on the sketch:

"Alexander defeats the Persian in a war.
Alexander burns the Pharisee palace and the Avesta to ashes.
The Persian runs away on a horse to the north.
The Pharisee runs away on foot to the south."

Esther discovers another piece. "What is this sketch? It looks like an *ashram*, the *rishi ashram* we visited."

The *swami* agrees. "Yes, it is the *ashram* a few thousand years ago. Alexander, the prince of Greece, was at the Indus River 2,400 years ago. The *ashram* has been a home for many successive *rishi*s since that time. The *rishi* you saw is the successor *rishi* at this time."

Christian says, "I know Greece is a part of Europe, and my parents came from Europe. I heard the Europeans are *Aryan*s, and I do not understand what that means."

The *swami* explains that the word *Aryan* in *Sanskrit* means noble, wise, or learned *Brahman.*

Christian, curious, wants to know if that means European languages are rooted in *Sanskrit.*

The *swami* affirms this. "*Manu* and his spoken languages are rooted in *Sanskrit*, and the written scripts are rooted in Brahmi."

Esther wants to know the difference between *Sanskrit* and Brahmi.

The *swami* clarifies it this way: "*Sanskrit* is the sound of vibration, and the sound is written in a pictorial script that is known as Brahmi, a left-to-right or right-to-left or up-and-down script."

Christian requests the *swami* to read the script written on the sketch.

The *swami* reads like this:

Alexander:
"I thank you, Brahman, for your welcome and blessings.
I heard about you from my teacher, Aristotle.
I am pleased to have come to see you in "Indoos."
I am grateful to receive your message of dharma.
I thank you for giving me the Veda and Sanskrit writings.
I will reintroduce this knowledge in the West.
I plan to build a library in Egypt and keep these writings."

Fig.6.6: Alexander at the *Ashram* with *Rishi*

The *swami* keeps reading from a note that reads:

"Alexander left with the Sanskrit writings.
Thereafter, Bharat became known as 'Indoos,' or India.
The Veda people became known as Hindu."

Christian finds another sketch. "It looks like *Buddha* teaching disciples. Is that right?"

The *swami* looks at it and agrees. "Yes, this appears to be one of the successor *Buddha*s that came a few hundred years after Alexander's time. The script reads like this."

Buddha:
"O disciples, go teach the message of righteous life.
Go to the West and to the East."

Fig.6.7: successor-*Buddha* with disciples Zeus, Nagarjuna in Himalayas

Esther thinks the *mantra* is the same. "Live righteous life doing a duty. That message comes from the Veda."

Christian stumbles onto another sketch. "What is this clip? Is there a history to learn?"

The *swami* reviews it and informs them that it is a conversation between Jesus and the Pharisee. "I will read it, and you decide if there is a history."

Pharisee: Where are you from?

Zeus: I am from here.

Fig.6.8: Zeus and Pharisee in a Pharisee-Temple

Pharisee: Who do you think you are?

Zeus: I am a son of the Father, God.

Pharisee: What are you doing here?

Zeus: I am serving the poor and sick.

Pharisee: Go and serve the sick. Do you see any sick here?

Zeus: Yes, I do.

Pharisee: Am I sick?

Zeus: You know the answer. One who thinks his own life is superior to all other life must be sick.

Pharisee: Who told you that?

Zeus: Father. You divide the sons of the Father by teaching the false.

Pharisee: I know what is right for my people. I make the rules. You go and get lost, or I will report you to the Romans. You are a troublemaker…

The *swami* reads a descriptive note written on the sketch, which reads like this:

"On the way out of the temple, Zeus creates a scene in the flea market.
He tells the poor not to buy the useless goods from the rich.
He tells them not to pay high prices.
The police come and take Zeus away.
After a trial, Zeus is nailed onto the cross."

The *swami*, looking at the next sketch, says, "Look at this sketch and see if this tells you any history."

Esther thinks there is much to learn from this sketch. "Jesus is nailed onto the cross, but I heard he came back to life after death. I always wondered how that could be possible. I never understood."

The *swami* thinks it could be true in two possible ways. "First, it could have been a made-up story to sell the Bible. Second, if Jesus was a master practitioner of *yoga nidra*, he would have felt no pain or spilled any blood on the cross and found wounds later, only after waking up in the cave."

Mohammad finds another sketch and gives it the *swami* to read. "What does this clip say?"

Fig.6.9: Zeus on the cross

The *swami* looks at it. "I will read it, and you ask me if you do not understand."

Fig.6.10: Pharisee and Paul in a Pharisee-Temple

The *swami* says, "This is a conversation between Pharisee and Paul, and it reads like this."

Pharisee:
I re-wrote portions of the Avesta into the Torah.
I will translate select portions of the Torah into Greek.
You know Greek is a left-to-right script, like Brahmi.
In the new book, I will make Zeus the son of god.

Paul:
What is the purpose?

Pharisee:
We lost power when the Persians lost to the Greeks.
The Greeks and the Romans hold the power now.
It is the future we must control, and that we can.

I devised a plan and it is easy to do.
All we need to do is sell the book in the name of Zeus.
We can be rich and powerful again, like before.

Paul:
It is a good idea, and free will always works in business.
It serves our purpose well.

Pharisee:
You know Greek, so be my partner in this business.

Paul:
Yes, I know Greek, but there is a problem.

Pharisee:
What is the problem?

Paul:
I do not believe Zeus is the only son of god.
The other day a gentile said so to belittle our varna.
You know what I did to the man? Kill, and ask no questions.

Pharisee:
That makes you a true believer, if you can convert.
Act in selling the book, and that will do it.
You may become a hero to the Greeks and Romans.
No one would ever know your past history.

Paul:
If you think that can work, I will give it a try.
I may have to create a story to make believe.

Pharisee:
Yes, you can do that.
You have free will.

The Hindu worships dark idols and a dark Krishna.
A white Zeus, in the likes of Krishna, makes it easy.
The paganu Greeks and Romans will fall for it.
They are a good target people to sell the book to and convert.

The *swami* looks at a note on the sketch and says, "The note reads like this."

"An old woman, another woman, and a boy disagree.
They persuade the Pharisee to act righteously.
The Pharisee would not listen to reason.
The Pharisee goes ahead with the trial, the cross, and the book."

Christian understands that all Pharisees are not the same. "That tells us some Pharisees act righteously. Why did this Pharisee act unrighteously or this way?"

The *swami* says, "It is *karma*. His *karma* must have made him insecure, unhappy, and blind with power; or else made him believe in Avesta, thinking the Veda is a mythology made up by brown-dark *manu* who lost senses in the hot *Surya*."

Esther wants to know if there are two groups within Veda followers, one that thinks *Vishnu* is all powerful and the other that thinks *Siva* is all powerful.

The *swami* says, "Yes, there are two groups, and they are aware that the *shakti* of *Vishnu* and *Siva* come from *mah-Isvar*. *Isvar* joins *Vishnu* and *Siva* in *Indra* and separates in *Surya*, all for a purpose. In that sense the *shakti* of *Isvar* is paramount."

Fig.6.11: Paul on his way to Damascus to sell the Bible

Mohammad finds another sketch with a horse. "What is it?"

The *swami* reviews the sketch and reads the text.

"Paul fell from the horse on his way to Damascus.
He is hurt and feels dazed.
He looks up at the shining Surya.
He imagines a blurred Zeus in the shining Surya.
He murmurs a prayer…"

Paul: "O Zeus, help me sell the book; I am a believer."

Mohammad finds a set of sketches in a clip and thinks they may reveal some history.

The *swami* looks at the sketches and agrees. "Yes, this clip has a set of three sketches that can provide a gist of history. It gives you a glimpse of Roman history some 1,690 years ago (325 CE). I will read the written words on each sketch, and you figure out what took place at that time. The first sketch reads like this."

"Roman Emperor Constantine was at war with a co-Roman emperor.
Constantine needs more people to join his army and fight.
Pharisee-Christian (310 CE) sees a chance to make a business deal.
Pharisee enlists the support of the queen mother of Constantine.
The queen mother, a converted Christian, negotiates a deal."

Constantine:
I want your help to get your people on my side.
I want to win the war.

Pharisee:
Yes, I know.
That's the reason I am here.
I can help you win this war if we can make a deal.
My soldiers will join your army and fight.

Constantine:
What is the deal?

Fig.6.12: *paganu*-Constantine and Pharisee deal making in palace

Pharisee:
First, you have to replace the swastika with a cross.
Second, after winning the war you must declare…
The Roman Empire is a Pharisee-Christian empire.

Constantine:
No, I cannot do that.
My people and I follow Veda.
We worship the sun god and the law of nature.

Pharisee:
Zeus, the son of god, and the cross can make you win.
Think it over and advise me.
I can do business when you are ready.

The *swami* looks at the second sketch and gives a summary that it appears the queen mother and Pharisee are negotiating a deal.

Fig.6.13: queen-Mother and Pharisee negotiate a deal in palace

The *swami* looks at the third sketch and reads the conversation aloud:

Mother:
Son, I think you should listen to Pharisee.
Zeus will make you win the war.

Constantine:
Pharisee is asking a big price that I cannot pay.
I do not know who this Zeus is.
I can't ask my people to give up Veda and the sun god.
We can't give up the 'Reality' and adopt a strange Zeus.

Mother:
I am going to suggest a compromise—
Give choice to the people, let them use free will.

Those willing can convert, adopt Zeus, and follow the Bible.
The rest can remain paganu and follow Veda, like you.

Fig.6.14: queen-Mother and son-Constantine in palace

Constantine:
All right, you make the deal with Pharisee.
I am a paganu and will worship the sun god until I die.

Esther finds another clip of two sketches and gives them to the
swami.

The *swami* reviews them. "In the first sketch the conversation
appears to be a deal in the making between the Pharisee and the
pope. I will read the clip and you figure out your own understanding
of it."

Fig.6.15: Pharisee and Pope in the Church of Zeus

Pharisee:

O Pope, you know you are my brother.

We did it together, the impossible.

We converted all pagans in the West.

They are now our soldiers.

They agree to follow our Bible.

We converted Isvar temple into the church of Zeus.

Now we must go on, no time to rest.

We must convert all non-Pharisees on the earth.

Make them believers in Zeus and follow the word of the Bible.

First, we should get the confused Persians and Arabs.

Then we should get the Hindu pagans in the East.

Pope:

Yes, I agree. It is a good idea.

The more we convert the more money and power to us.

Tell me how to do that, an easy way.

Pharisee:
The quick and easy way is to make war, defeat, and convert.

Pope:
I agree. It is the best idea I ever heard.

Pharisee:
We shall begin the war in Zeus's birthplace…
And go on to conquer all lands.

Pope:
I agree, let us move.

The *swami* reads the notes at the end of the above conversation.
"The Bible cult (325 CE) in its first wave goes to war with the Arabs and Persians and starts converting the defeated to follow the Bible, the Bible that came out of the ashes of Avesta.
An Arab-Persian, Mohammad, forms a Koran cult (620 CE) and starts converting the defeated to follow the Koran, the Koran that came out of the ashes of Avesta. The Koran cult declares jihad as 'Allahu Akbar.' Thus the war in the name of false gods continues."

The *swami* reads the notes on the second sketch.
"The Koran cult, armed with swords, destroys ancient temples of gods/goddesses. And it converts the Isvar temple in Mecca as Allah-masjid. It continues the war and takes over the right-to-left-script people and their lands in the Persian Empire, Indus to Turk, Turk to Egypt, Egypt to Ethiopia in Africa. It marches on into India, destroys ancient culture, writings, and forces the Veda people to adopt Allah. It converts an ancient Pharisee temple in the Kashmir Valley into Allah-masjid."

Fig.6.16: Pharisee Christian v. Muslim Mohammad,
'Bismillah Allahu Akbar'

"The Bible cult starts a second front. It takes over the Veda
people and lands in the Americas, India, and in between and
makes them follow the Bible. Then, the Pharisee-cult forms a
Red Book cult in Russia…Thus the three cults, the Bible, Koran,
and Red Book, have been busy ever since in converting the rest
of the Veda people as if there is no tomorrow for the false gods."

The *swami* inquires if this part of ancient history is clear.

Christian responds, "Yes, sir, it is clear. The ancient Pharisee
temple is in the Kashmir Valley. That must be an interesting site
to see. I found this sketch of the *rishi*. Please tell us what it says."

The *swami* looks at the sketch and agrees to read it aloud:

"The *rishi* at the *ashram* is the one who succeeded the prior *rishi*. Such
successions have been going on for millions of years. The present *rishi*,

being at the end of his time, is sitting before the *murti* meditating until the deer-*manu* arrives. Finally the deer-*manu* walks into the *ashram*."

The *swami* reads the conversation between the *rishi* and the deer-*manu*, which goes like this:

Rishi:
Swagatham

Deer-manu:
Namaste, maha-Rishi
I, as a deer, used to be a disciple at this *ashram* many years ago.

Rishi:
Namaste
Yes, I know. You are a manu in this birth.

Deer-manu:
I came back here to receive your blessings in this life.

Rishi:
Yes, you have my blessings to do your righteous duty.
I pray for *Isvar* to keep you in the path of dharma.
I know you came here to do your duty.
I know you want to tell me more.

Deer-manu:
Yes, as a deer I listened to the Veda and Bhagavad Gita.
I always wished to find an answer to the question—
Where is Isvara?

Rishi:
I know you discovered the 'Truth.'
Therefore it is your dharma to be a successor rishi.

Fig.6.17: successor-*Rishi* and Deer-*manu* in the *Ashram*

"Rishi closes eyes, as if his righteous duty has come to an end. And since the successor has arrived, the rishi stops breathing."

Christian thinks that is the reason the present *rishi* came to be known as "dear *rishi*." He says, "Thank you, sir. We have to go a long way to get to the Kashmir Valley."

Glossary:

Yoga nidra=Deep sleep

Avesta=A perverted-Version of *Sanskrit* sounds of Veda written in right to left script of Brahmi

Asura=Nature of a living being doing righteous duty for selfish purpose

Sura=Nature of a living being doing righteous duty for self-purpose

paganu=Living being that worships nature

Chapter: 7

vignaanam yagnatanute | karmaaNi tanutemvicha |
Vignaanam devaahasarve | brahma jyeshthamupasate |"

Knowledge spreads the feast of sacrifice
Knowledge spreads the feast of works (deeds)
Gods offer adorations to Him as to Brahman, the 'Elder' of the Universe.
(*Taittiriya Upanishad*)

Adventure for knowledge

Esther, Mohammad, and Christian arrive next in the historic Kashmir Valley. On their way to an historic ancient masjid, they walk into an antiques shop. An old man in the shop greets them.

Fig.7.1: Esther, Mohammad, and Christian with an old-man in a shop

The man says, "Come in. I have everything you want before you visit the masjid."

Mohammad says, "We come from America. We are students visiting Kashmir to learn about the old culture."

The man tells him, "I know you look familiar. Were you here before—or maybe not? I can help you. Ask me anything you want to know. In fact, a rumor has been going around here for years, as long as I can remember, that an ancient relic is buried under this Pharisee temple. Now it is a masjid, so go study and learn. Do not touch or break anything. We had a professor from America who wanted to dig under a wall for some bones. The guards did not allow him to do that. The professor took a few photographs and left. The guards are strict and protect the masjid. It holds relics that are much older than those at our holy Mecca."

Mohammad says, "If the guards are so strict we may not be able to carefully study the ancient relics."

"You can persuade the guards to cooperate," the man tells him.

Esther asks, "How do we do that?"

The man says, "Do them a small favor and bring a gift or two."

Mohammad asks, "Can we get those gifts here?"

"Yes, you can."

The old man packs two items for them and whispers, "This one, give it to the guards when you see them at the main door. There are usually two guards at the door. The guards work a twelve-hour shift. Night shift begins at six in the evening and ends at six in the morning. In the night shift, they smoke a lot to keep awake, and not many visitors show up. This one, keep it with you and put a small quantity in the smoke-pot, if needed. We like it here; it makes you feel good."

"What is it?" Esther asks.

"It comes from a plant. It is good to smoke," the old man says.

Mohammad says, "Can I also get a small knife, a small hammer, a piece of rope, and a piece of adhesive tape."

"Here it is," the man says, "all you want. The total you pay, a student price, is one hundred dollars."

Mohammad says, "I do not have that kind of money."

"How much you have?" the old man asks.

Mohammad says, "Ten dollars."

"Give me that," the man says. "It's a discount price for student visitors. Be good in the masjid and make no trouble. The guards have guns, and they shoot first and ask questions later."

Christian says, "Thanks. We will be good, no trouble at all."

Fig.7.2: Christian, Mohammad, and Esther at Masjid entrance

The three visitors arrive at the entrance to the masjid. They see two guards standing at the entrance with rifles in hand. Mohammad gives the gift to the first guard who begins to search their bags. The guard takes the gift and lets them go in while looking at Esther.

The second guard says, "Enjoy your visit and stay as long as you want. Read the instructions, and do not touch or break anything. Also, do not be scared of the skeletons or other things you find. Remember not to go into any room that is closed. You can read and take notes using a flash if required. We will be watching you."

Mohammad says, "Thanks."

Fig.7.3: Christian, Mohammad, and Esther in a Masjid room

They begin to search the room with a flashlight and notice a hidden door behind a skeleton. Suddenly the first guard appears in the room.

"Did you find what you are looking for?" he asks.

Esther, shook up, says quickly, "No, we are not looking for nor have we found anything interesting yet."

The guard says, "OK, I will show you something that might interest you, so follow me. You two keep looking here until I return. Let us go, *bachi*."

The guard takes Esther into another large room. There, the second guard is smoking a pipe while sitting on a floor cushion. Mohammad follows them to help Esther if required.

Fig.7.4: Esther and guard dance as the 2nd guard smokes the pipe

The first guard says, "I went to school in America. I like dancing with girls. Let us dance and give a try at having fun. Your friends can keep looking for what they want."

Esther says, "OK, no problem."

The guard and Esther start dancing. The second guard continues to smoke his pipe on the floor cushion. Meanwhile Mohammad watches the scene from behind a door. While dancing, Esther reaches into her pocket and pulls out a small bag that looks like a tea bag and quickly drops it into the smoke-pot, which is standing on the floor behind her.

After some time, the second guard starts dancing with Esther while the first guard smokes the pipe. That routine continues until the guards are intoxicated and fall asleep on the floor. Esther's dropping of the drug bag into the pot did the job. Mohammad comes in, ties up the guards' hands and feet, and puts the tape over their mouths. Then Esther and Mohammad go back and join Christian in the search.

Christian tells them that he has located the hidden door behind the skeleton but that the skeleton has to be moved before they can see if the door will open. To avoid attracting the guards' attention, he has not tried moving the skeleton. Now the three of them together try to move the skeleton away from the door, but they stumble and turn the skeleton sideways, which opens the hidden door. Behind it is a dark chamber that appears empty save for a large picture hanging on the wall.

When they tap the wall close to the picture, the three sense a hollow wall behind it. They carefully pry loose a few bricks from the wall using their hammer. Placing his hand in the wall cavity, Christian feels a small box. He pulls it out and opens it to find a small pack of cards. They keep the box and put the bricks back in, closing the wall opening. They come out of the chamber, turn the skeleton to its original position, close the chamber door, and walk out of the masjid quietly, without looking back.

Next, they return to the *ashram,* where the *rishi* studies the pack of cards found in the box.

The *rishi* compliments them on their adventure. "You did your righteous duty. This is a *TaaLapatra* and it is a part of the Veda, not the complete Veda. This part gives you the details as to how *Indra* controls *surya-bhoomi-chandra.* Thus it is the foundation for the rest of the Vedas."

Christian say, "*maha-Rishi,* permit me to ask you a question. We also discovered that your name, 'dear *Rishi,*' is actually 'deer *Rishi.*'

I understand that many years ago, a deer disciple at this *ashram* listened to the Veda and *Bhagavad Gita* as taught by a *rishi* of that time and later was born as a *manu* and came back to this *ashram* as a successor *rishi.* Please tell us how that came about."

"It is true," the *rishi* agrees, "but it is the deer *and* the calf, not the deer alone. The two disciples at the *ashram* were reborn later as male/female humans first. And their child is the deer *manu* who came here to become the deer *rishi. Isvar* creates one *shakti* by joining the half-*shakti* of sperm with the half-*shakti* of an egg."

Fig.7.5: Christian, *Rishi*, Esther, and Mohammad with Veda in the box

Esther asserts that this is DNA's twenty-three chromosomes of genes from sperm and twenty-three chromosomes from the egg. "Is it not?"

The *rishi* prefers to use simple words to make the concept clear, so he explains it this way: "The DNA, or chromosomes or genes, is a *paramanuvu* cloud, a code of atoms linked together, whose density varies in each individual in a specific/particular way, consistent with prior *karma*. The code by itself will not function alone until the *shakti* activates it. Therefore DNA or genes can be a rough road map from where *Isvara*, the *atmam*, can start creating or destroying. Thus *atman* drives the code to assemble the required *anuvu* and creates an individual living being with certain abilities."

Esther reinforces the concept of *shakti* by pointing out the problems faced by scientists. "In biology the code is known as deoxyribonucleic acid, DNA, or genes. But the scientists are clueless as to what drives the code. May I ask how the deer or calf became human?"

The *rishi* explains, "By listening to the Vedas, the deer and calf acquired the *jnana*, which purified the *atman* by getting rid of the excessive 'cloud' in the *paramanuvu* in which *atman* resides."

Christian says, "If I understand it correctly, the righteous duty of the deer and calf caused a decrease of the density in the '*paramanuvu* cloud,' which modified the chromosomes in the cells. And that change is the cause for improved ability in the next birth. In science, I learned that when man modifies the chromosomes in a seed, it improves its ability to produce a better crop. Do you think these two processes are similar?"

"Yes," the *rishi* agrees. "But the reverse can also be true. An unrighteous *manu* can acquire more 'cloud,' lose ability, and be reborn as an animal."

Christian wants to know the summary message that one should keep in mind at all times.

The *rishi* says, "The Almighty *Brahman* is One, and we live within to do a duty, our righteous duty, for the *param-brahman*, not for a selfish purpose. *mah-Isvar* is within the *param-brahman*, and a small part of *Isvar*, known as *Isvara* or *atma* (consciousness or self or life or soul), is within each living individual. This is to be purified in a birth-death cycle, or cycles, to merge back with *mah-Isvar* for the use of the *param-brahman*. The *param-brahman* is the reality, like each of us, and *Isvar* is the reality that runs the actions within the *param-brahman*, and *Isvara* runs the actions within each living individual. *Manu* cannot know the size or shape of the *param-brahman* being within. Similarly, the little living creatures within *manu* can't know the size or shape of the *manu*."

Christian wonders if there is only one Almighty *Brahman* in whom we live or if there may be more in the same manner that there are many *manu*.

The *rishi* says, "This is a big question. I do not know how to answer. It is possible there are many *param-brahmans*."

Christian follows up with a question. "If there are many Almighty *Brahmans*, then is *Isvar* within all the *param-brahmans*?"

"True," the *rishi* agrees, "and *mah-Isvar* is eternal."

Glossary:

bachi=girl
DNA=deoxyribose nucleic acid or genes: an embedded code in the cell

Chapter: 8

aksharam brahma paramam

Brahman is that which is immutable and
independent of any cause but Itself
(*Bhagavad Gita*)

aneka vaktra nayanam, anekaadbhuta darshanam
aneka divyaabharanam, divyaanekodyataayudham
divyamaalyaambaradharam, divyagandhaanulepanam
sarvaascharyamayam devam, anantam viswato mukham

Krishna displays gigantic personality of *Brahman*
with innumerable faces on four sides...
with fragrance of heavenly perfumes, and innumerable arms....
(*Bhagavad Gita*)

The Reality

The School of Physics Lecture
Topic: The Structure of our Universe and the Motion of Matter

Moderator: Physics Chair
Panel: Professors Christian, Esther, and Mohammad
Audience: Students, Professors, and Scientists

The moderator welcomes all to the lecture. "Today we will hear
from three professors on their findings on reality and the adventure
that began nearly sixty years ago with a visit to India, the land of
mystery. Reality is still unfolding today as we gather here. I present

you with Professors Christian, Esther, and Mohammad, and I invite them to discuss their findings with you. And you are free to ask questions at any stage during this presentation."

Christian welcomes all and thanks them for being here today. "I will give you a summary on our discovery. In short, we, as students, visited India during one summer some sixty years ago. We did not know it was going to be that hot up there. We landed in an airport and saw the newsflash on TV and in a newspaper, which said 'Dear *Rishi* saw *Isvara*.' It was as though the news reported 'Dear Newton saw Gravity.'

I did not understand the news. So I inquired about it to someone in the airport and learned 'dear' is 'deer'; '*rishi*' is a '*yogi*'; and '*Isvara*' is the 'Creator,' which roughly translates into 'a deer in a previous life is a *yogi* and saw the Creator'; interesting news indeed. So we took off to see the *rishi*. The *rishi* lived in a small hut in a remote valley with wild tigers as his neighbors."

Esther welcomes everyone and continues, "The hut is an *ashram*, and when we got there, it was filled with people. They were singing songs evidently in the names of 'god,' and we sat on the floor with them. After the prayer, *Rishi* welcomed us and encouraged us to ask questions, so we did."

Mohammad welcomes everyone and thanks them all for being here. "I learned some things for the first time. Initially I did not want to go there thinking it would be a waste of time. I admit it turned out to be the most enlightening experience in my life. The *rishi* was humble and noble and had vast knowledge about reality."

Christian continues and goes into some details. "I learned that we, all living beings on the earth, live within an 'Atom' of the Almighty god. The *shakti* of *mah-Isvar*, which is present within the Almighty, creates the sun, stars, planets, the earth and the atoms, you and me, and the rest of the living beings from the atoms. The *shakti* of *mah-Isvar* creates, destroys, and recreates all the living beings in our universe for a specific purpose using the DNA code embedded in *Indra*, a Neutron. The ancients knew of this knowledge millions of years ago, or before the last ice age, and the knowledge of old *Sanskrit* sounds, which is enunciated in the Veda and are written on *TaaLapatra* in old Brahmi script."

Mohammad adds more detail. "We found a part of that *TaaLapatra* in Kashmir, buried in a wall of a masjid. The masjid is an ancient Pharisee temple under the control, now, of the Muslims."

Esther explains the principle involving cold and hot force. "We took that *TaaLapatra* back to the *rishi*, its rightful owner. The *rishi* explained to us about the principle behind the *shakti* existing within cold and hot matter:

- The intensity of force at a distance R from a cold or hot emitter = 1/R
- The resultant intensity of (cold plus hot) force at an emitter = 1/(RxR)
- If we use a space constant, B, the resultant intensity of force= B/R^2.

Newton utilized this principle in deducing gravity equations, but he did not identify the source of the principle. Here, we show how *Indra* controls all the objects in an Atom including the sun, earth, and moon."

Christian explains how he has modified Newton's equations applying the principle of cold plus hot force. "We modified the equations deduced by Newton using the principle of *shakti*. And using the modified equations, we deduced the structure of the Atom in which we live. We discovered *Indra* is the center of the Atom, while the sun is *not*. *Indra* orbits and rotates the sun; and the sun in turn makes Earth and other planets rotate.

Indra and the sun together make Earth and other planets go around the sun in noncircular orbits. *Indra* makes Earth, planets, moons, and other objects orbit in an inclined plane, not at 90 degrees to the *Indra*-sun line. *Indra*-sun also makes Earth's axis of rotation a ZZ wobble spin. The variable gravity and dynamic weather on the surface of the earth is due to its rotation, orbit, and spin. The Atom is like an elongated sphere or ball of fluid having a variable diameter of approximately 722AU, with submerged and floating suns, planets, moons, ring-objects, comets, and asteroids, etc. orbiting around *Indra*."

Esther goes into some specifics and explains the details. "See Figures 8.1 and 8.2 that show an Atom in which we live. Figure 8.1 shows our sun and the star on line at 180 degrees to *Indra* or on the opposite sides of *Indra*. The view on the XZ plane, looking at YY, shows *Indra*, the star, sun, the earth, and moon—the rest of the planets are not shown for clarity. The large circles around the sun and star represent the heliospheres, a mix of hot and cold matter. The arms 108AU and 238AU across are the force bands that keep the sun and star orbiting *Indra* at those distances. Here, the distance between the sun and Earth is an astronomical unit, AU=1.496×10^{11}m.

The *swastika* with its legs pointing to the left, shown at the center of *Indra*, depicts the direction *Indra* rotates as well as the orbit direction of the cold matter, sun and star. The *swastika* with its legs pointing to the left, shown in the heliospheres, at approximately 90 degrees to the plane on which *Indra* rotates, depicts the direction the sun and star rotate and the direction of the hot-matter orbits."

Esther continues to explain. "Figure 8.2 shows the present position of the sun with respect to the star and *Indra*. Here, the sun is at approximately 90 degrees to the star with respect to the center of *Indra*. The sun is moving closer to the star on the same side of *Indra*. I am sure you have questions on these figures, so let us discuss."

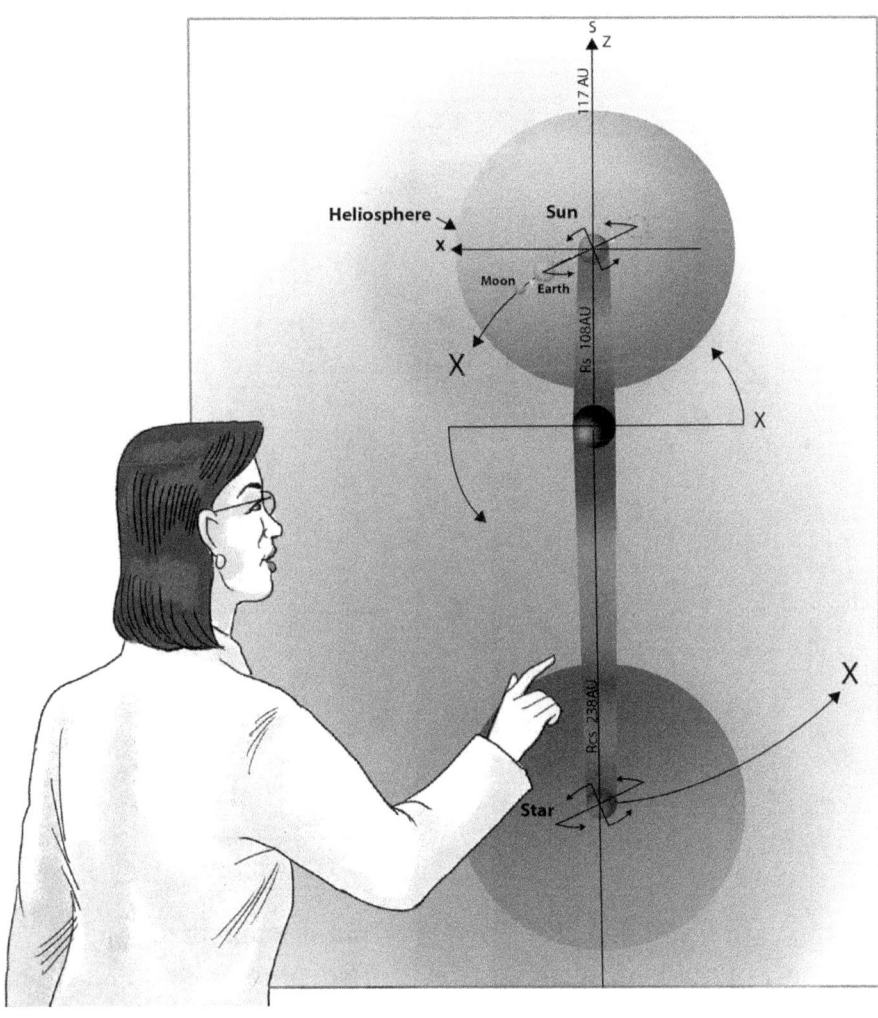

Fig.8.1: Esther explains Atom with Sun-Star on opposite side of *Indra*

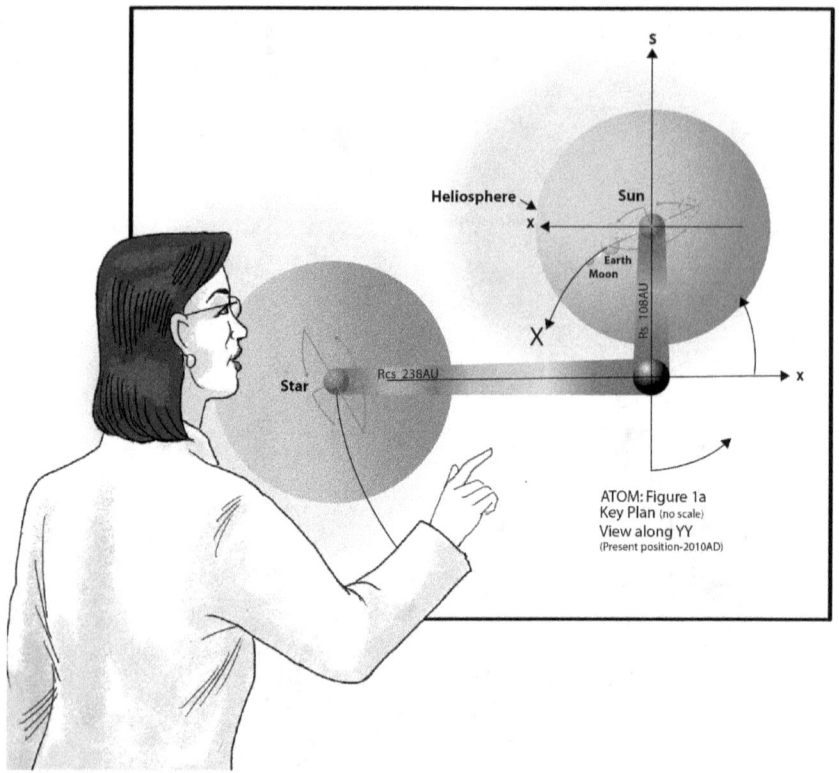

Fig.8.2: Esther explains Atom with Sun-Star moving to the same side of *Indra* (The position at the present time, ~2010CE)

A voice in the audience yells: "The findings and the proof in support of the Vedas is not novel, nor equal to seeing the action with your own eyes. Nobel does not recognize this sort of work."

"I am Copernicus," says another speaker. "Not the real one. I am a great-great-great-grandson. I have a simple question. Are you telling us in so many words that my great-great-great-grandfather was wrong and the sun is not the center of our universe?"

Esther says, "Yes, I am sorry."

Another speaker says, "I am Galileo; not the real one, a descendant. May I say you must know the real Galileo cannot be wrong because billions of us, including scientists, agree that Galileo is correct, and we see the sun as the center in our universe. I do not know where you were hiding when Galileo lost his life advocating what is evident. I hope you are not making this up with a daydream for an alternative after everything has been discovered and settled."

Esther disagrees, saying clearly that Galileo and the billions of others, including the scientists, are wrong.

A voice in the audience says "I am professor Stone from Caltech, and I am the chief scientist on NASA's Voyager 1 & 2 sent into space ~36 years ago to explore the space—they are still alive and active. They explored the nose of the heliosphere at ~100AU from our-sun; now, the Voyager 1 left the heliosphere and it is more than 125AU away from our sun—it is expected to go on… and on. So, my question is- where is your *Indra*- Voyager 1 did not run into it at 108AU?"

Esther answers "Thank you professor Stone for your comment and the inquiry. First of all Voyager could not have gone towards the nose of the heliosphere—going to the nose would be against the flow. For example, see which way the satellites go—they mostly in the direction the earth rotate and orbit. Likewise, the Voyager must have gone in the direction the sun rotates and orbits. That meant Voyager had gone to the tail-end of the heliosphere, not the nose end—it is still alive and active at the tail because, it is a low temperature region. The reason it did not run into *Indra* at 108 or 125AU must be the same, sun or the earth or the satellites did not run into *Indra* either—i.e. the Voyager plane of orbit is close to Earth plane of orbit. If we re-route Voyager path and direct it to head towards *Indra*, it may get there one day."

Another speaker says, "My name is Pope. I am not the real one; the real pope is my hero. I heard all about Galileo fighting with the pope. Now we know that the pope knew all along that the sun is not the center of our universe. Do you agree with the pope at last?"

Esther says, "Yes, I agree the sun is not the center of our universe, but Earth is not either. So both Galileo and the pope were wrong."

Pope says, "The real pope can never be wrong. The pope talks to Jesus all the time, and Jesus knew it all. Jesus sits on the right side of the Father and listens to the Father and tells everything to the pope."

Another speaker says, "I am Jihad Mohammad, not the real one, although I come from the same root and follow the teachings. Call me Jihad. I know Jesus is a prophet and there is no Father. It is only 'Allahu the Akbar,' whoever that might be. Given that, the pope must be talking to himself, because Allah would never speak to the infidels. Oh little pope," he says to the other man, "tell me if Earth is flat."

Pope says, "Oh Jihad, I know Earth is not flat."

Jihad says, "If that is true, do you agree your pope is wrong to claim that Earth is flat?"

"No, I do not agree," Pope says. "If you go out and look, it looks flat. My pope meant that kind of flat."

Jihad says, "That proves your pope is talking to himself, not to Allah."

Esther says, "That kind of arguing is not going to move us forward. Please ask a question and do not argue."

Another speaker says, "I am Newton, a descendant of the famous Isaac Newton. Please tell me how you deduced that *Indra* controls two suns."

Esther says, "The National Aeronautics and Space Administration (NASA) has collected data on the sun and planets, such as the distance between Earth and the sun, orbit-velocity, and the mass, etc. We can verify that observed data only if we use:

- *Indra* mass   =4x10^30kg,
- Sun to *Indra* distance   =108AU or 108x1.496x10^11m,
- And, per NASA, sun mass =1.989x10^30kg.

If the mass of *Indra* has to be twice that of our sun, then there must be a second sun in the Atom. Per Chadwick (1930 CE) neutron mass in an atom equals to 1.006 times the mass of the protons. So *Indra*, a Neutron, should have a mass equal to the mass of two suns, which equals 1.006(2x1.989x10^30kg)=4x10^30kg."

Newton asks, "How did you determine the distance between *Indra* and the second sun?"

Esther says, "We know of the helium atom (He) and its physical or chemical properties. For example, it has two protons and two electrons, and we know through experiments how much force (F) or energy (E) is required to pull the electrons away from the neutron in the atom He. We can apply this principle to estimate the amount of force that would be required if someone wanted to pull Earth away from the sun and *Indra*. The force (F) or energy (E) ratio of the two electrons in atom He equals E1/E2 ~2.2.

That difference in the force (or energy) between the two electrons in the helium atom must be due to the difference in the orbital distances of the electrons from the neutron. Since there is no change in the mass of the protons or the electrons, the force ratio must be equal to the distance ratio.

So take the force ratio equal to the distance ratio: the distance ratio, dE2/dE1=E1/E2 ~2.2; or, the distance between neutron to proton2 / neutron to proton1~2.2.

Likewise, if someone wants to pull the sun away from *Indra*, the distance ratio must be identical in the Atom—or in our universe; *Indra* to the second sun's distance is approximately 2.2(108AU)=238AU."

Newton asks, "How do you know both suns orbit in the X direction?"

Esther says, "We know all planets orbit in the same direction and approximately in the same plane, the XY plane. See in Figure 8.1, the sun-star system shown on the XZ plane with the Y axis at 90-degrees to that XZ plane. If *Indra* rotates anti-clockwise about YY in the XZ plane in the X direction, then the sun and star must orbit in the X direction, the way the planets orbit in the direction that the sun rotates."

Newton inquires, "As we all know, the real Newton did not know what makes Sun's force transfer to Earth or how that force controls Earth's orbit. So my question is- do you know how the forces transfer between Sun and Earth or *Indra* and Sun in making Earth or Sun orbit?"

Esther answers, "Let us see Figures 8.1 and 8.2 carefully. The legs of the *swastika* at *Indra* point toward the X axis, and the legs show the direction *Indra* rotates and the cold matter orbits. And the legs of the *swastika* at Sun point toward the Y axis, and the legs show the direction Sun rotates—in a plane at 90 degrees to the plane in which *Indra* rotates, and the legs show the direction the hot matter orbits. *Indra*'s cold matter orbiting in X direction gets entangled with Sun's hot matter orbiting in a plane at 90 degrees. That entangled cold-hot matter creates a force which makes Sun orbit in X direction and rotate in Y direction. Similarly, the entanglement of Sun's hot matter and Earth's cold (+hot) matter makes Earth orbit and rotate; and the entanglement of *Indra*'s cold matter and Earth's hot (+cold) matter makes Earth axis of rotation ZZ wobble-spin which causes the pole-reversal in 4.32x10^6 year cycles."

"Wait a minute professor. My name is Reeves and I am a director of research on Cosmos at Los Alamos National Laboratory. Recently NASA put two satellites in space to probe Van Allen belts; and the latest observations use unique and right instruments that span a broad range of energies with amazing sensitivity in Earth's equatorial orbit that cuts through the belts at different altitudes. From that data we can resolve whether something is changing in time or in space or both. As you may be aware the Van Allen radiation belts are two concentric, doughnut-shaped rings that are made up of high-energy electrons that vary in intensity. The belts are confined within the Earth's magnetosphere and extend from about 1000 to 60,000 km above Earth's surface. Although discovered more than 50 years ago, the Van Allen belts are not yet fully understood— previous observational data of 1990s did not fit the theories, and the researchers are confused as to what processes really control the intensity of the radiation belts. The new data shows electromagnetic waves in the solar wind are almost certainly transferring their energy to the trapped highly energetic charged particles, electrons and protons, within the belts boosting them to local and radial accelerations. So my question is- how does your theory fit into my findings?"

Christian takes the question and answers this way. "Dr. Reeves, I thank you for the update on your latest observational evidence and the findings. First of all the concept of doughnut-shaped rings is not accurate although it appears that way if the observed data is only at or near Earth's equatorial orbit. When you make more observations above and below Earth's equatorial plane you will see the shape of the rings or the magnetic flow lines that look more like number 8—the flow lines between N-magnetic pole and the equator slope downward, and between equator and S-magnetic pole slope upward. Similarly, Sun's hot matter flow lines go from N-magnetic pole to the S-magnetic pole. The acceleration of the particles is real in local and radial directions, but not in situ. As pointed out before, the entanglement of Sun's hot matter with Earth's cold (+hot) matter produces a 'moving force band' between Sun and Earth to keep Earth in its proper orbit and rotation; the 'force band moves forward' as Sun rotates and the entangled-atoms get split up and accelerate away. If local and radial accelerations are observed for these atoms it must be due to the release angles of these entangled atoms—it is an atom not an electron."

Newton asks, "What is the orbit period or the time the sun takes to go around *Indra* once?"

Esther says, "It takes about 700 years for our sun to orbit once around *Indra*. And it takes about 2,500 years for the companion sun to orbit once around *Indra*."

Newton says, "We know planets orbit at different speeds. My question, do you know if the sun and star orbit at different speeds?"

Esther says, "Yes, we know our sun orbits at an average speed of about 1.5×10^{11}m per year, and the companion sun orbits at an average speed of about 0.9×10^{11}m per year, or at 60 percent speed of our sun, which is close to half the speed of the sun."

Newton says, "As the planets do, at some point in time the two suns come closer and go farther apart in their orbits. Am I correct?"

Esther nods. "Yes, they come closer on the same side of *Indra* once in about every 1,000 years. At the present time, they are moving closer and the sun is at about 90-degrees with respect to star-*Indra* line. In the next 200 years or so, both suns will line up on the same side of *Indra* at approximately 130AU apart. Then, about 500 years thereafter, the suns will be on the opposite side of *Indra* at about 346AU apart."

"I am Kepler," says a voice, "a follower of Johannes Kepler's laws. Do you know if the orbits of the sun and the star trace ellipses?"

Esther says, "We have no way of knowing about that at this stage because we have no data to either prove or disprove this. Kepler had the orbit data on Mars, collected by Brahe, from which Kepler determined that the planets trace ellipses. Likewise, we need the observed data on the orbit of our sun before we can determine whether the orbit traces an ellipse. If I have to take a guess, I believe Kepler's law is universal whether or not the orbit itself traces a true ellipse. It is so because two half-*shakti*s create and control the one *shakti* whether it is Mars or the sun or within a living being. As a result, the one *shakti* will have its own unique features, as does Mars or other planets, and traces an ellipse with variable speeds when acted upon by two half forces, in our case the sun and *Indra*. Similarly, two *Indra*s may be acting upon the sun, assuming the Atom is a part of a molecule. That may be the reason for the 7.5 degrees of tilt in the ZZ axis of our sun. See Fig. 8.4. Therefore I will not be surprised if I find data in the future that shows the orbit of our sun traces an ellipse."

Newton says, "I am glad to hear that. As you know, the real Kepler or real Newton did not know why a planet orbits that way, tracing an ellipse. What would be the impact on Earth during this thousand-year cycle when the two suns are on the same side of *Indra*, or on the opposite side of *Indra*?"

Esther says, "I agree that Kepler and Newton did not know of the existence of *Indra*; no other scientist did either. Also, note that the mathematics or calculus claimed to have been deduced by Newton some 330 years ago (1680 CE) had its origin in the Veda, the knowledge of which Arya Bhatta (500 CE), his disciple Bhaskara I (630 CE), and later Bhaskara II (1150 CE) introduced to the world hundreds of years before Newton's time.

When the two suns come on the same side of *Indra* at 130AU apart, once in a thousand-year cycle, the temperature on Earth will rise and the climate will be hot with tornados, cyclones, and asteroids coming toward Earth. When the two suns move to the opposite side of *Indra* at 346AU apart, once in the thousand-year cycle, the temperature on Earth will drop and the climate will be ice cold with snowstorms."

Pope asks, "How does that impact the living beings on Earth?"

Esther says, "The living beings suffer loss of life and destruction of property during the hot climate due to tornados, cyclones, and from possible asteroid hits. They will suffer damage during the cold climate due to the snowstorms."

"The Bible reveals exactly that," Pope says.

Jihad says, "No, the Bible reveals the past history on wars and crimes, not about the future."

Newton says, "In Figures 8.1 and 8.2, the intensity of shade in the heliospheres decreases from the right side to the left side. Why?"

Esther says, "If we call the right side of the heliosphere on the X axis the nose, the left side of the heliosphere becomes the tail end. The legs of the darker *swastika* indicate the direction in which *Indra* rotates and the cold matter orbits. The legs of the lighter *swastika* indicate the direction in which the sun rotates and the hot matter orbits. The sun rotates on the XY plane or at approximately 90 degrees to the XZ plane, the plane on which *Indra* rotates or the cold matter orbits.

And looking at the heliosphere to the right side of our sun, or at the nose, we can see that the hot and cold matter move in opposite directions at a greater speed, especially the cold matter, so the collisions cause a rise in the temperature. On the left side of our sun, at the tail end, the cold matter speed gets reduced by the time it reaches the tail end due to continued encounters of the hot and cold matter all along its route, from the nose to the tail end, so the collisions occurring at a lesser speed cause the temperature to decrease at the tail end."

Newton asks, "What makes the sun rotate?"

Esther says, "Per NASA, our sun's axis of rotation at ZZ tilts about 7.5-degrees. That could mean the sun's ZZ axis wobbles as it spins, like Earth's ZZ axis. If in fact the sun's ZZ axis has tilted 7.5 degrees with respect to *Indra*'s equator plane of XZ, as shown in Figure 8.4, such a tilt would create a force couple on the sun's surface, which acts through the force band between *Indra* and the sun. And that force couple acting on the sun's surface on the XY plane rotates the sun anti-clockwise around its ZZ axis looking from *Indra*. In other words, *Indra*'s X-force (FisX), acting at the center of the sun, makes the sun orbit while the difference in *Indra*'s X-force (FisXt) at the top and (FisXb) at the bottom of the sun, acting upon the sun's magnetic field flow, rotates the sun anti-clockwise. Or the net effect of the force band between *Indra* and the sun would create a force couple at the surface of the sun due to the fact that *Indra*'s X-force at the top of the sun would be greater than the X-force at the bottom of the sun."

Kepler asks, "Do you know if the YY axis of *Indra* has a wobble spin?"

Esther says, "It is possible, but I have no knowledge of that at this point. However, as discussed before, *Indra* being one *shakti* within the Almighty *Brahman*, *Indra* may have been created by two half-*shakti*s, like DNA with its 23 + 23=46 chromosomes. If so, the creator of *Indra* may be the controller, and that makes *Indra* orbit, rotate, wobble, and spin, like Earth."

Jihad says, "It is Allah that creates, not the *Brahman*."

Pope says, "It is Jesus that creates; no one else can."

Another speaker says, "I am a rabbi, a follower of rabbi Zoroaster, the *asura* god who gave us the Avesta, the root of the Torah, Bible, and Koran. I say this: Let us not fight about the name of the god after following the *asura* god for thousands of years. Let us agree to renew our loyalty to the 'god of Abraham and Moses,' whoever that might be, the only god I know."

Another speaker says, "I am a *Vishnu* follower, *Vishnu*, the creator who sleeps on the snake god Cobra in *Indra-loka* (heaven) and created the universe. He came down onto the *bhoomi* in the form of the god *Krishna* and taught *manu* the way of living. I say this: *Vishnu* is *Krishna*, and *Krishna* is the *Brahman*. There is nothing else; the rest is all a *maya*."

Another speaker says, "I am a *Siva* follower. *Siva* created the universe, came down onto the *bhoomi* in the form of the god of tribes, and taught *manu* the way of living. I say this: *Siva* is *mah-Isvar* and dances in the Himalaya hills with the snake god Cobra in the neck and creates, destroys, and re-creates all the living beings for a purpose. There is nothing else; the rest is all a *maya*."

Christian makes a point to the *Vishnu* and *Siva* followers about reality. "Years ago, I learned from the *rishi* that *Vishnu* and *Siva* join in *Indra* to be a part of *mah-Isvar*. In fact, my own research shows reality is *mah-Isvar*, *param-brahman*, and *dharma*. There is nothing else."

A person in the audience inquires, "I am a follower of Adi Shankara's *Adwaita*—non duality. That tells us we are all One within *Brahman*, and the *Brahman* is *Isvara*. You sound like preaching *Dwaita*—duality. *Isvara* and *Brahman* are not One. Please explain why?"

Christian clarifies, "In science we ask—is it the wave or particle. The wave you can see and a particle you can't, although the wave doesn't exist without a particle. Similarly, we can ask—is it the *Brahman* or *Isvar*. The *Brahman* you can see and *Isvar* you can't, although the *Brahman* doesn't exist without *Isvar*. That way *Adwaita* and *Dwaita* meant the same."

"I am Oppenheimer," says another speaker, "and an expert in genetics. Based on my study of the genes in man and ape bones, I conclude that there is a 95 percent chance that man evolved from an ape. That makes it probable that humans from Africa migrated into Israel first, a nearest education center in the caves and desert. From Israel caves, man must have migrated into various places including present-day India. There, the native tribes took the migrant leader with lighter blue skin as the god *Krishna*, thinking he must have come down onto the earth from heaven; later, the tribes made *Krishna* a Jesus, the son of god. Thus all the people of the other eleven tribes on this earth worship the Pharisee that originated in Israel caves. Therefore there is no mystery. The god of Abraham who made the ape to evolve into Adam and Eve prevails."

The rabbi says, "Well said, and that is the reason we must believe in the word of the chosen."

"I am Yuliner," says another speaker, "and I wanted to be a scientist but ended up as a Wall Street gambler. Yes, I am one of the chosen and believe in the 'god of Abraham.' My god made me rich and powerful. Be ready for the surprise of your life. One day, I and my gambling buddies Smoky, maker of the smoke that keeps you high, and Gunny, maker of the guns to make you go low, will take over the earth and become the 'god' of all."

"I am professor Incandela, a scientist," says another speaker. "First, I want to thank Yuliner and his gambling buddies for sharing with us a few bucks out of the hard-earned billions in gambling with the lives of little people. Second, I want to report on an important finding that I and more than 3,000 other scientists have seen in an underground experiment. Believe it or not, we saw a 'god-like' particle. We could do that only because of the good-luck bucks of Yuliner. Of course, we get billions from the European Union, although the EU is going into a deep black hole. This work should continue forever because I need to figure out more precisely the existence of this 'god,' whoever that might be. Yuliner should keep gambling and keep giving us money to do more experiments until we are able to see the 'god' of all."

Esther says, "Thank you, professor. I am glad you saw the 'god-like' particle. Tell us more about your experiment."

Incandela explains in simple words to make everyone understand. "Simply put, this is how it works: We took a can of hydrogen atoms (H), stripped the electrons off of the atoms, and uploaded the protons, a bunch at a time every few seconds, into an accelerator. The accelerator incrementally increases the speed of each bunch of protons close to the speed of light, $c = 3 \times 10^8$ m/s. And the superconducting magnets direct the jet to crash into another jet of protons coming in the opposite direction. The sensors record data on the particles created in this crash. A summary of that data consistently showed a 'bump' or a spark at $125\times10^9 eV$. That unique spark appears in my eyes to be a 'god-like' particle."

Jihad has a simple question. "The caveman saw the sparks too, rubbing two stones or hitting one with the other. What is the difference here?"

Incandela gives some thought before answering the question. "I must say there is no comparison; it is like day and night. In my experiments, two jets of protons come at each other at the speed of light, collide, and break up into their constituents displaying sparks at a variety of energies. I picked a spark, which is 'god-like.' The spark generated by the caveman, even if you can call that an experiment, is very simplistic and had no sensors to measure the spark; it made no sense or there is no god-like particle."

Esther asks, "What about the fire the caveman discovered?"

Incandela answers, "That could have been discovered by anyone, but not the god-like particle."

Esther asks, "Did the caveman discover the 'fire god'?

Incandela says, "Sure, you can call that a fire god for lack of a better word. My discovery is more fundamental in that it goes into the fire god."

Esther says, "When you stripped off the electron from the hydrogen atom, what did you do with the neutron? When you made each bunch of protons speed up in a given direction using superconducting magnets, what happened to the electromagnetic particles that caused the protons to speed up?"

Incandela gives some thought to his answer. "First, the hydrogen atom has no neutron, only a proton and an electron; second, the issue of electromagnetic particles is a minor problem compared to the particles created by the proton-proton crash."

Christian gives Incandela an update. "I must disagree with you, sir. The proton or electron cannot survive without a neutron in an atom, even in the hydrogen atom. Allow me to ask a simple question. Is it possible the energy bump at $125 \times 10^9 eV$ you saw might have been the product of 125 protons crashing into each other at that instant?"

Incandela agrees. "Yes, it is possible, because each proton contains energy equal to approximately $1 \times 10^9 eV$."

Jihad says, "So it is not a 'god particle.'"

Christian asks, "Professor Incandela is it possible that each spark at an energy level might represent the crash of a certain number of protons at an instant?"

Incandela nods. "Yes, the proton jets come from opposite directions and crash. I have no way of knowing how many protons in a bunch crash at any given instant."

Christian goes deep into the issue and raises another question. "I understand that a given proton in a jet goes at a particular speed and at an angle with respect to the center of the jet, and it would crash into either a proton or a bunch of protons coming from the opposite direction at a different speed and angle. Do you think that would give you a clue why you saw sparks at a variety of energy levels?"

Incandela says, "It is possible, but I have no clue."

Another speaker says, "I am Marx, a follower of the great Karl Marx, and I say this to the gamblers and the believers in god: Wait a minute, we Marxists are coming. We are nearly two billion strong, and before you know it, we will take over the earth by joining hands with two billion Islamists. I too come from the family of the chosen by the 'god of Abraham,' whoever that might be. However, I do not believe in 'god.' The 'god' crowd is confused and lost. Marx is my god, and let all people be converted into Marxism and live happily ever after."

Esther says, "I assume no one has any questions to ask?"

Newton has one simple question. "What makes *Indra* rotate?"

Esther says, "It is a difficult question to answer. I think it is the *shakti* of *mah-Isvar*, which is within the Cell of the *param-brahman* or the galaxy of the Universe."

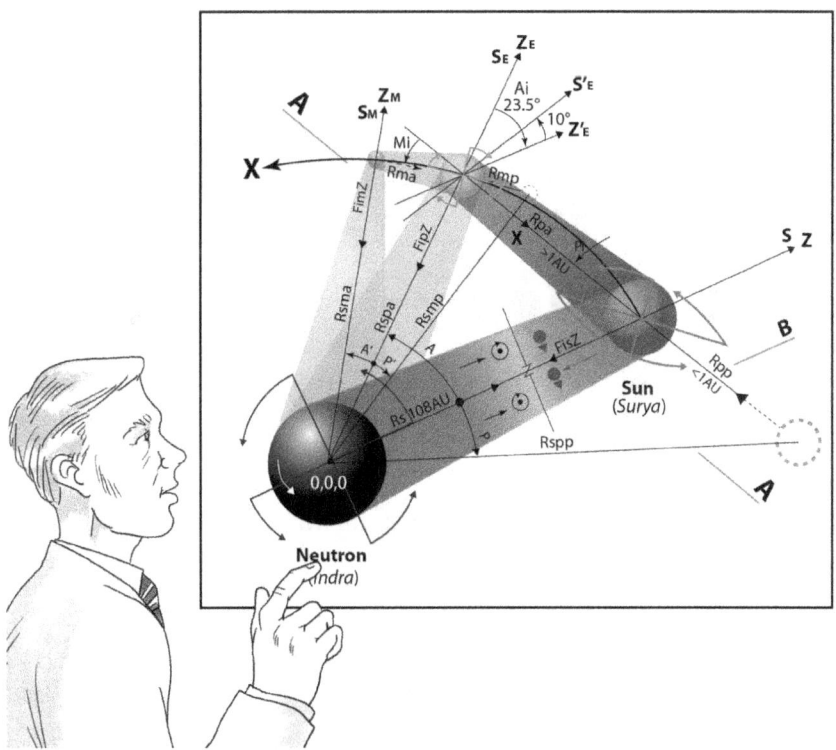

Fig.8.3: Christian explains Atom with *Indra*-Sun-Earth-Moon

Christian explains the specific details of what makes the earth orbit, rotate, and spin. "Now let us look at Figure 8.3. It is the Atom in which we live. It is a view on the XZ plane looking at the YY axis. It shows *Indra* with the sun, earth, moon, and five force bands. The sun, under *Indra*'s X-force (FisX), orbits in the X direction, and under the force couple of *Indra*'s X-force (FisX), the sun rotates about its ZZ axis. The five force bands keep the sun, earth, and moon in proper motion as they orbit, rotate, and spin. The bands exist between *Indra*-sun, sun-earth, earth-moon, *Indra*-earth, and *Indra*-moon."

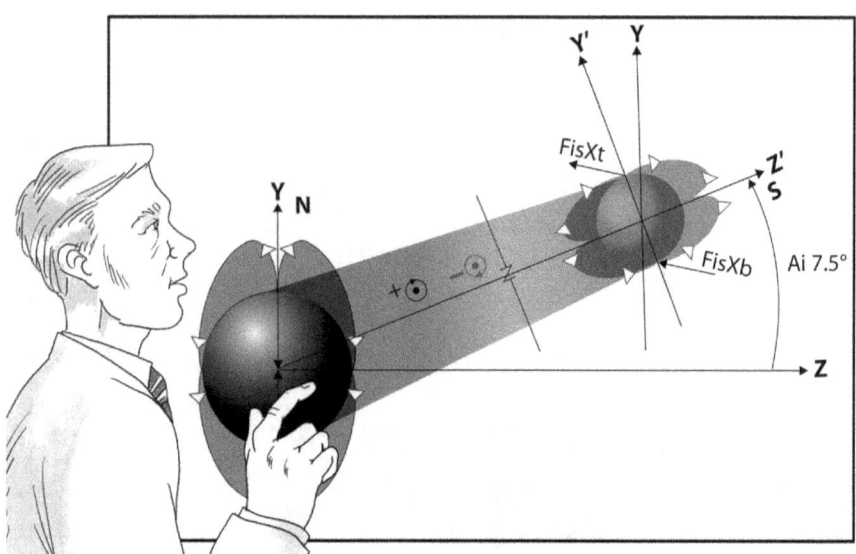

Fig.8.4: Christian explains *Indra* cold-field flow, and Sun hot-field flow

Christian continues to explain Figure 8.4. "It is the view <u>BB</u> on the YZ plane looking along the XX axis. It shows *Indra*'s cold-field flow, the sun's hot-field flow, and the force band between *Indra* and the sun. Due to the 7.5-degree tilt of the sun's axis of rotation of *ZZ*, *Indra*'s X-force (FisXt) at the top of the sun's surface would be greater than the X-force (FisXb) at the bottom. That difference would create a force couple on the surface of the sun that, acting on the surface of the sun, rotates the sun anti-clockwise, or in the direction of the legs of the *swastika* at the sun."

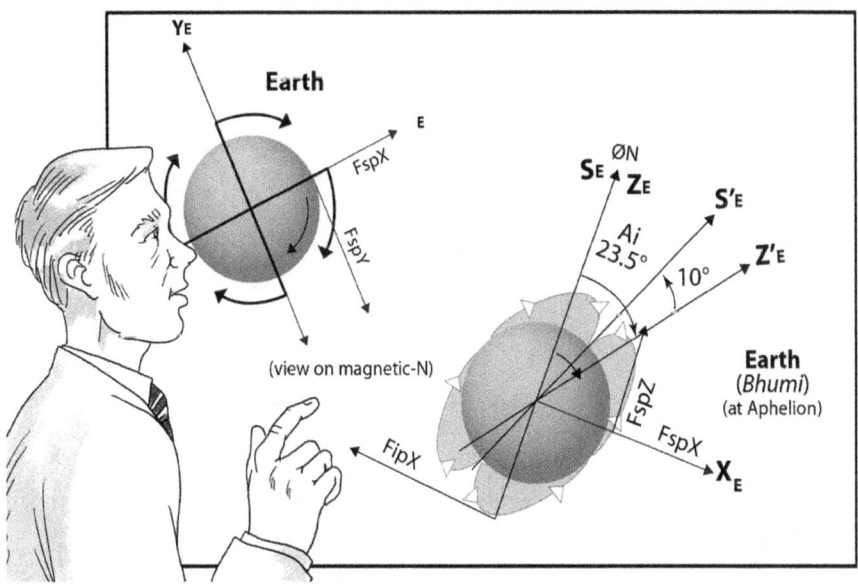

Fig.8.5: Christian explains Earth magnetic-field flow

Christian continues to explain Figure 8.5. "It is a view on the XZ plane looking along the YY axis. It shows Earth with its magnetic field flow, *Indra*'s X-force (FipX) acting clockwise upon Earth's field flow while the sun's Z force (FspZ) acts anti-clockwise upon Earth's field flow. The interplay between the clockwise force, FipX, and the anti-clockwise force, FspZ, upon Earth's field flow results in a net clockwise force acting on Earth's surface, which causes Earth's axis of rotation ZZ to spin clockwise continuously at a slow and varied speed."

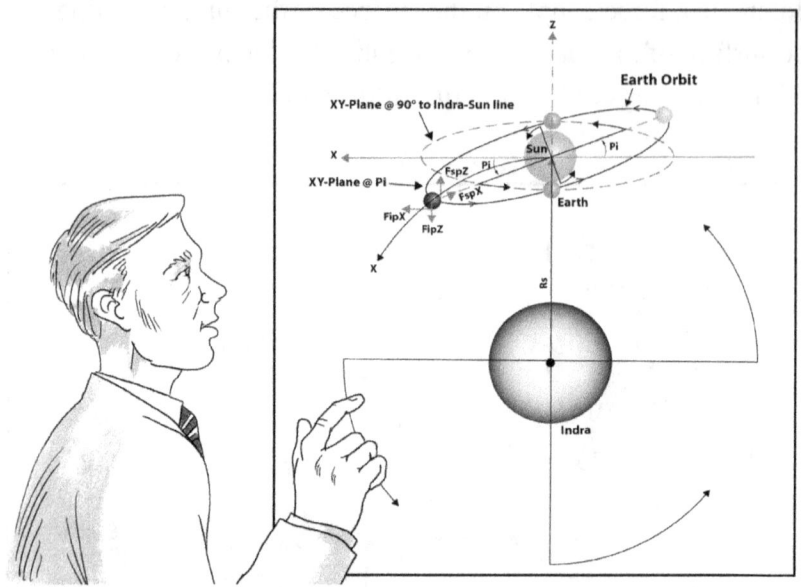

Fig.8.6: A view of *swastika*s at *Indra* and Sun with Earth's orbit, Christian explains Earth's orbit-inclination to XY-plane caused by *Indra*'s Z-force (FiZ)

Christian explains Figure 8.6 and shows what makes the earth spin. "Now let us find what makes the earth orbit, rotate, and spin. This is somewhat difficult to visualize. Figure 8.6 is a view from the XY plane looking along the ZZ axis from *Indra*. We know *Indra*'s X-force (FisX) makes the sun orbit and rotate. And the calculations (see Chapter 9) show that the sun rotates once in twenty-seven days, which confirms NASA's observed data. As the sun rotates anti-clockwise on the XY plane, it creates a Y-force (FspY) on the earth, and that FspY-force acting on the earth's field flow allows the earth to orbit anti-clockwise on the XY plane (looking from *Indra*) once in 365 days. The earth rotates clockwise, or in the direction the legs of the *swastika* at the earth (looking from *Indra*), once in approximately twenty-four hours. I will stop here, and let us discuss your questions."

A speaker says, "I am Einstein; not the genius Albert Einstein, but I came from the same family. I can clearly see that you have come up with an alternative to that great man's work. Do you know the spin period or spin speed of the earth?"

Christian answers "yes" and continues to explain. "We spent a lot of time on that question because it is not easy to find the correct answer. First, we analyzed IERS data (International Earth Rotation and Reference Systems Service). Based on that data and the present given tilt angle of the earth at $Ai\sim23.5E$, we estimate the present rate of spin of the earth's ZZ axis at $Zs23.5-0.2$ arcsec per year.

We know Earth's field flow intensity (see Fig. 8.5) comes out high at the magnetic north, and it follows along a downward curved path to the equator, where the flow decreases to low and becomes horizontal. Then the flow follows an upward curved path to the magnetic south, where the flow increases to high and turns into the magnetic south. Given that, we needed to deduce an equation to describe the interplay of the forces that cause Earth to spin under *Indra*'s X-force, the sun's Z-force, and Earth's field flow, the solution of which should give us $Zs23.5=\sim0.2$ arcsec per year.

Using a trial-error method, we deduced an equation with a constant, C. The equation we deduced, which we presented in Chapter 9, gives us an exact solution. The exact solution yields $Zs23.5\sim0.195$ arcsec per year, and this result from the equation exactly matches with the values given in *Bhagavad Gita* (Veda). The rate of spin or spin speed that varies each year and the values obtained from the equation match the criteria specified in the Veda with respect to *yuga* and its four unequal time periods. The summary of the results are as follows: We divided the spin into four 90-degree segments starting at 0N (North Pole). Clockwise from 0N: 0-90, 90-180, 180-270, and 270-360 (0). Then we subdivided 0N-180S into four phases as:

Phase-3, 0N-45E (180S-225W)

Phase-4, 45E-90E (225W-270W)

Phase-1, 90E-135E (270W-315W)

Phase-2, 135E-180S (315W-0N)

And, for our purpose we assumed-

Indra X-force acts at 180S (in X-direction, clockwise).

Sun Z-force acts at 90E (in Z-direction, anti-clockwise).

We designate 0N-45E as Ai=0-45 and so on..."

Table: The result obtained using the deduced-equation.

Ai (Phase 3)	Rate of spin-speed	Period*
0	0	0
45E	0.375 arcsec/Yr.	0.864x10^6 years
(Phase 4)		
45E	0.375	
90E	0.375	0.432x10^6
(Phase 1)		
90E	0	0
135E	0.1875	1.728x10^6
(Phase 2)		
135E	0.1875	
167E	0.0965	0.811x10^6
180S	0.0965	0.485x10^6

*The total period for 180 degree spin equals to 4.32x10^6 years as given in *Veda*.

Christian continues to explain. "At the present time with Earth's ZZ axis tilting to Ai~23.5E, we are about three-quarters of the way into phase three. And at the end of phase four, Earth's North Pole (magnetic south) faces the sun at Ai=90E. Thereafter, Earth continues to orbit the sun showing the same face toward the sun for thousands of years. During that time Earth's magnetic south and magnetic north poles flip, once every 182 days or at perihelion and aphelion in every orbit, from one hemisphere to the other until the pole reversal occurs. The pole reversal means Earth's magnetic south at aphelion, which is facing the sun at Ai=90E, turns into magnetic north. And that will be the beginning of a new *yuga* in phase one (Ai=90E-135E). So the spin period of Earth in one full cycle from 0N back to 0N, which is 360 degrees, would take two *yuga*s, which equal 8.64x10^6 years. This is a phenomenon similar to an electron orbit; the electron in its orbit spins parallel, anti-parallel, and back to parallel."

"I am Professor Mitrovica," says another speaker. "I study and measure changes in the orientation of the earth's magnetic field stored in ancient rocks millions of years ago; thus, I am able to see the effects of the oscillatory polar wander using a computer model of the true polar wander. Polar wander is the relative movement between the earth's surface and its axis. The earth's surface tips over and the pole shifts up to 50 degrees and then turns around to its original position; this takes about 10x10^6 years. Now you are telling me it takes 8.64x10^6 years. So here is my question: Do you agree that my study of ancient rocks and the model are more accurate than your equations or the mythology and philosophy of Veda?"

Christian thanks the professor for his study of ancient rocks and the discovery that the pole shifts up to 50 degrees and then turns back to its original position and to do so takes 10x10^6 years. "It appears the pole took 10x10^6 years of time to shift 50 degrees plus 50 degrees equals 100 degrees. My work shows that the shifting pole would not stop at 50 degrees and turn back. Instead the pole continues to shift and makes a full circle of 360 degrees in 8.64x10^6 years. So there is nothing much to agree with your work, and there is no reason for the pole to shift up to 50 degrees and decide to turn back."

Einstein wonders what the impact would be on living beings from such a spin of the earth.

Christian explains, "Earth's spin would cause extreme climate changes, and in phase four it could cause the end of all living beings on Earth. By the end of phase four, Earth's North Pole would spin and face the sun (at Ai=90E). Then the poles keep reversing to magnetic south at aphelion* to magnetic north at perihelion* and that process continues for thousands of years. In other words, Earth's magnetic north at perihelion reverses back to magnetic south at aphelion. Thus Earth could orbit the sun for thousands of years showing the same face to the sun as the moon does now per NASA. During this period, tens of thousands of years, Earth's wobble spin and rotation would come close to zero, or Earth may rotate slowly once in each orbit under *Indra*'s Z-force (FipZ). Thus the hemisphere facing the sun would have hot-age with light at the polar region, and the other hemisphere would have ice-age with dark at the polar region. And the equatorial region would have no visible light. When *Indra*'s X-force (FipX) begins to spin Earth's magnetic south at aphelion, it turns into magnetic north at aphelion, a pole reversal, and it moves forward beyond 90E."

Mohammad says, "I am going to speak about the moon's motion and what causes it to orbit, rotate, and spin. Per NASA, the moon orbits the earth at an orbit inclination (Mi) of approximately 5 degrees. Some outer moons of Jupiter and Saturn have an orbit inclination greater than 75 degrees.

---

*Aphelion = The point in Earth's orbit closer to *Indra* or farther from the Sun, and
*Perihelion= The point in Earth's orbit farther to *Indra* or closer to the Sun, in each orbit.

Our moon orbits the earth about once a month showing the same face to the earth. Per NASA, the 'moon is in synchronous rotation about the earth.' The moon takes as long to rotate on its axis as it does to orbit once around the earth. That way, the moon always keeps the same hemisphere facing the earth."

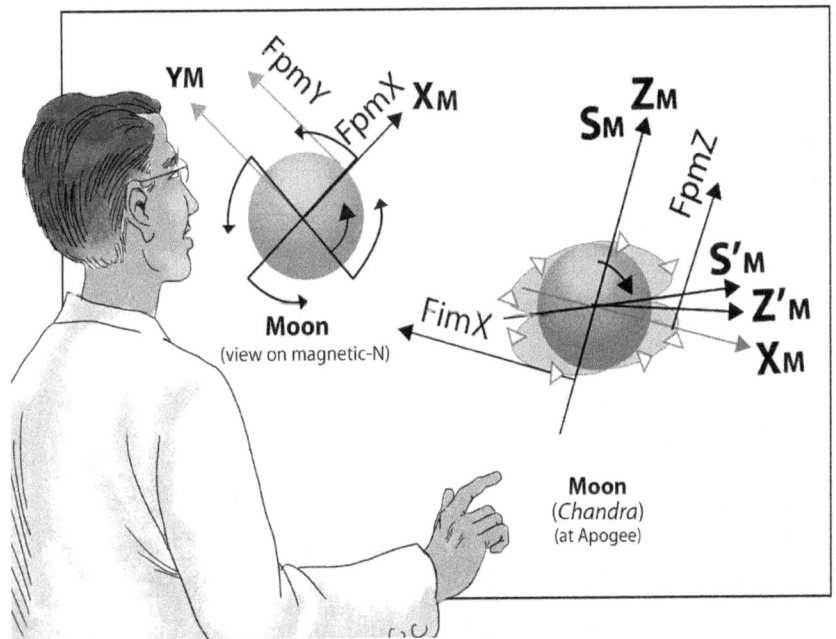

Fig.8.7: Mohammad explains Moon with magnetic field-flow

Mohammad continues to explain, "In the case of the earth, we discussed that it could keep orbiting the sun at Ai=90E for thousands of years facing the same hemisphere to the sun and that it could rotate slowly or once in each orbit. Here, the present status of the moon (see Fig. 8.7) appears to be orbiting Earth showing the same hemisphere and its south–north poles flipping approximately once every fifteen days.

The earth's Y-force (FpmY) that acts on the moon causes it to orbit around the earth in an anti-clockwise direction, or opposite the earth's orbit, and rotates the moon clockwise. *Indra*'s X-force (FimX) and Earth's Z-force (FpmZ) act upon the moon's field flow, and the interplay of *Indra*'s force and Earth's force upon the moon's field flow causes the moon's ZZ axis to spin clockwise toward 90E to face the earth. And at 90E, the moon could keep orbiting Earth with the same hemisphere turned toward Earth for thousands of years, as it does now per NASA.

With respect to the outer moon orbit inclination of greater than 75 degrees or the retrograde/polar orbits, that could occur when the distance between the planet and the moon (Rma) increases. An increase in Rma would increase *Indra*'s Z-force (FimZ) on the moon. Such an increase in the force would result in *Indra* pulling the moon closer, thus causing the moon's orbit plane angle to increase. Such an increase in the angle creates a polar or retrograde orbit for the outer moon. Please feel free to ask questions."

A voice says, "I am de Sitter, a member of the clan of the famous Willem de Sitter. I know my hero de Sitter discovered that the sun and the earth control the motions of the moon. And he knew there was no invisible hand in space that could make the objects move the way they do. Of course, other than the famous Einstein's curvature of space-time and the general relativity (GR) equations that can solve every conceivable problem in space. That being the fundamental prevailing principle over the past hundred years or so, may I ask how you can say an invisible hand in space is doing all these acts, not Einstein's GR? How can you tell me de Sitter is wrong and the sun has no control over the moon?"

Christian wants to answer this issue. "I agree it is hard to dispute the findings of the great de Sitter, knowing of his collaborative efforts with the great Einstein in such discoveries. However, we have no choice but to tell you about our findings that the sun has no direct control over the motions of the moon.

It is the earth and *Indra* that control the motions of the moon, and that is true of all the satellites that go around the earth. *Indra* is the invisible hand in space; therefore Einstein's GR is not the answer to the orbit inclination of the planets or the moons, including retrograde orbits."

de Sitter says, "If sun-*Indra* controls the earth, and earth-*Indra* controls the moon, what affect would other planets have on the sun, earth, or moon?"

Christian says, "The simple answer is: none directly. With the planets being so far away they do not affect the motion of the sun or the earth or the moon. They mix the hot and cold matter in space and help create the atoms."

de Sitter says, "If the sun is orbiting around *Indra*, then it is not the center of our universe as Galileo, Kepler, and Newton thought."

Christian agrees. "Yes, *Indra* is the center of our universe."

de Sitter asks, "What is *Indra*?"

Christian explains, "*Indra* is a Neutron in the cold, and it must be the nucleus with the genetic code, a deoxyribonucleic acid (DNA), of the Atom. *Indra* is the provider of the material to build the sun (star), planets, moons, and other objects in our universe, including you and me. To build in its own image, destroy, and rebuild is a process that goes on for a purpose."

A voice says, "I am a Bohr and related to the genius Niels Bohr who discovered the atom. I have a question. Per Bohr, an atom has a nucleus consisting of a neutron plus a proton with an electron orbiting the nucleus. Are you telling us that the great man was wrong?"

Christian answers, "Yes, the Bohr atom and the devised electron shell structure make no sense."

"I am Oppenheimer," says another speaker, "a follower of the great bomb maker Robert Oppenheimer. I have a question. If the proton is not held by the neutron in an atom's nucleus, where is the strong force in a bomb?"

Christian says, "There is no strong force other than the force of gravity within the neutron."

Oppenheimer says, "Do you really mean there is no other force besides gravity?"

Christian says, "Yes, I mean it and I know it."

Einstein says, "Oh wait a minute, I heard you dismantle GR, the atom's nucleus, and now the strong force. I wonder, what is going on with your newfound theory? Tell me, what do you think of fission and the fusion of an atom?"

Christian explains, "The fission of an artificial atom, created by man, splits its neutron into smaller atoms; fusion joins two or more neutrons into a larger atom. It is the neutron that undergoes fission or fusion, and there is no nucleus called the neutron plus a proton."

"I am Hubble," says a voice, "and the real Edwin Hubble was my great-great-grandfather. I have a question. Is there a difference between fission of an atom and gamma ray bursts (GRBs) in space?"

Christian continues to explain, "In the fission of an atom, the scientist adds a new neutron to an atom and splits it, creating a force for use. The GRB takes place in space or within a living being and is an act of nature in destroying star(s) in space or proton(s) in a living being; it occurs all the time as part of making or remaking an Atom or atom for a specific purpose. Per an ancient *rishi*, it is the '*Om*,' the creative vibration of the *shakti*, *mah-Isvar* within the *param-brahman*, or *Isvara* within a living being."

Hubble says, "As you may know, the real Hubble informed us that he had observed such GRBs in space with a telescope sitting on the earth, and his observations showed that the light from such a GRB recedes as if the distant galaxy is running away from us. Observing such events,

Hubble thought the universe is expanding. I ask you if Hubble was correct."

Esther says, "No, Hubble was incorrect. For example, let us say that our sun explodes tomorrow and a real Hubble sitting on an earth in another galaxy turns his telescope to the GRB after a few days to a month and keeps recording the intensity of the light flux. Guess what the record would disclose? Surely a diminishing intensity of light flux, and that does not mean our galaxy is running away from Hubble. Actually, that decrease in the intensity of the light flux may be for other reasons, including burnout of the fuel in the sun."

"Wait a minute, Professor," says another voice. "I am Schmidt, and recently three of us, two experts and I, spent a lot of time observing the phenomena and discovered that the galaxies are accelerating away from us at the speed of light, not just expanding as Hubble thought. In fact, on account of this fundamental discovery, we have received Nobel Prizes. I ask you to show me we are wrong, how we are wrong, and how you are correct."

Esther explains it this way: "I am sure you can find the answer in my response to Hubble. I will clarify it further. We know our earth rotates and orbits. As stated earlier, our sun orbits too. Likewise, the distant star must have orbited before and after it exploded. At the time Hubble and/or you observed the GRB, during a short period of a few months, our sun with the earth and the earth itself must have continued to orbit; plus the GRB must have continued to move in its orbit with a diminishing intensity of light flux by the minute. The intensity of the light flux diminishes due to (1) exhaustion of fuel in the exploded star, and (2) an increase of the relative distance between our earth and the exploded star. For that reason, your observed data and its use must be erroneous."

Schmidt says, "May I ask you another simple question? You conclude that the companion star is at 238AU from Earth, but the nearest star is 4.2 light years away from Earth. So where is your star?"

Esther answers this way: "The nearest star is the companion to our sun. The distance of 4.2 light years must be wrong. Your parallax as well as the red-shift or blue-shift method to measure distances in space is erroneous due to the presence of *Indra* and *paramanuvu*. As you and I know, the space in our neighborhood is filled with double stars. So astronomers are mistaken to count our companion star as part of a nearby double star and wrongly name that a three-star system. There are no three-star systems in space, not at least in our neighborhood."

"I am Lamaître," says a voice, "from the younger generation of Georges Lamaître. Are you telling me there is no big bang, no black hole, no expanding/accelerating universe, and that 'god' is everywhere?"

Christian affirms with, "Yes, yes, yes, and yes."

"I am Schwarzschild," says a voice, "from the family of Karl Schwarzschild. I do not believe a word of what you have said about the black hole. They do exist at the center of all galaxies.

Ask Professor Stephen Hawking of Cambridge or any other famous professors or an editor of a science journal anywhere on this earth. All of them would agree the universe would fall apart without black holes. So I ask you if you have invented *Indra* just to ridicule the existence of the black hole that the great god Einstein with GR predicted must exist."

Christian explains, "*Indra* is a Neutron. In 1930, James Chadwick discovered the neutron in an atom, and no one knows how it looks or what it does in an atom. We have discovered a Neutron in the Atom in which we live, and we are trying to show you what this Neutron is doing to our sun, our earth, and all living beings. If you and the rest think there is a black hole at the center of our galaxy, our work did not reach there yet. So there is no live dispute on that issue at this time."

The chair says, "How do you explain Nobel Prizes for the work on black holes, the accelerating universe, etc.?"

Christian says, "The Nobel Prize committee has to explain. I sent messages (see appendix) to the Nobel on this very issue."

Einstein says, "You say there was no big bang. Then how do you explain a cosmic microwave background (CMB) of 2.7K?"

Christian says, "The CMB is the leftover radiation of the sun/stars at *Indra* and is the radiation in our Atom at a distance of 108AU from our sun."

Einstein asks, "What do you think has happened to the radiation at the sun of ~6,000K or at corona ~1×10^{6}K?"

Christian says, "The *paramanuvu* in space absorbs/emits the radiation; thus it gets disbursed in space between the sun and *Indra* and beyond. And it acts as the communication signal link with other Atoms in the Galaxy or Cell."

The Chair asks, "Do you think the companion sun has anything to do with the climate change?"

Esther says, "Yes, certainly with respect to the rise in the temperature on Earth in a short time scale of ~250 years, or the destruction that will happen during the next 200- years, in a thousand-year cycle during which time our-sun and the companion-star would be on the same side of *Indra*. The intensity of rise in the temperature increases in each cycle due to the increase of the spin of Earth's ZZ-axis. If we start with our sun and the companion on opposite sides of *Indra*, it takes about a thousand years before the sun and the companion-star come back to that position again. That is a thousand-year cycle.

If our sun begins the cycle being in the cold and keeps orbiting *Indra* anti-clockwise at a speed of about 1.5×10^{11}m per year, during the same one year the companion sun orbits *Indra* anti-clockwise at a speed of 0.9×10^{11}m per year, being at a greater distance from *Indra* at about 2.2 times the distance between *Indra* and our sun.

That means if the companion sun were at the same distance from *Indra* as our sun is it would have moved forward a distance $(0.9 \times 10^{11})/2.2 = 0.41 \times 10^{11}$m per year. We know if the companion does not move, our sun will catch up with the companion in each orbit in about 700 years.

When the companion is also moving forward at 0.41×10^{11}m per year, our sun needs to travel a greater distance before it can catch up to the companion:

The extra time our sun needs is about $[(0.41 \times 10^{11}) \times 700]/(1.5 \times 10^{11}) = 191$ years. But during those 191 years the companion keeps going forward, so our sun needs more time, which may be about $[0.41 \times 10^{11}) \times 191]/(1.5 \times 10^{11}) = 54$ years and so on. When we add up all these years, it will give us ~1,000-year cycle. During this 1,000-year cycle, in each quarter of the time our sun moves: (1) from a cold-warm dark matter area into a warm-hot dark matter area, as at the present time, and comes closer to the companion on the same side of *Indra*, (2) from a hot-warm dark matter area into a warm-cold dark matter area, (3) from a warm-cold dark matter area into a cold-ice dark matter area, and (4) from an ice-cold dark matter area into a cold-warm dark matter area—then, the moves continue to repeat (1)-(4). During the present cycle, our sun will be in a warm-hot area for one-quarter of the cycle, or 250 years. So, we can expect a warmer to hot climate on the earth in the next ~200 years as the sun and companion star move to the same side of *Indra* at about 130AU apart."

A voice from the audience says "I am Dr. Pachauri, Chair of UNO's IPCC. The Intergovernmental Panel on Climate Change (IPCC) is the leading international body for the assessment of climate change. It was established by the United Nations Environment Programme (UNEP) and the World Meteorological Organization (WMO) to provide the world with a clear scientific view on the current state of knowledge in climate change and its potential environmental and socio-economic impacts. We review and assess the most recent scientific, technical and socio-economic information produced worldwide relevant to the understanding of climate change. Review is an essential part of the IPCC process, to ensure an objective and complete assessment of current information. Our work is policy-relevant and yet policy-neutral, never policy-prescriptive. So, on the basis of the scientific data available, IPCC reports that CO_2 is the primary cause of Climate Change or rise in temperature on the earth. Therefore, 'man is the primary cause—made warming on the earth extremely likely 95% and as the concentration of the greenhouse gases increase warming of the atmosphere and ocean continues, the snow and ice diminishes, and the global mean sea level rises.' Now, I hear you say 'the primary cause is the nature—the companion-star and the spin of Earth's axis of rotation'. Please tell me what is going on?"

"Professor Christian, before you answer Dr. Pachauri may I state my latest findings on climate change based on a model study at University of Colorado, Boulder. I am Professor White of Institute of Arctic and Alpine Research. In May 2013 the concentration of carbon dioxide (CO2) climbed to 400 ppm for the first time in modern history. The last time CO2 in the atmosphere reached 400 ppm was 3 to 5 million years ago during the Pliocene; and Earth was about 3.5 to 9F degrees warmer than it is today—Seas swelled pushing levels 65 to 80 feet higher. When we put 400 ppm CO2 into a model, we don't get as warm a planet as we see when we look at the records from the Pliocene. That tells us that there may be something missing in the climate models. When we tried a simple model without sea ice all year round, we got the right kind of temperature change and we got a dampened seasonal cycle, both of which are things we think we see in the Pliocene. Basically, when you take away the sea ice, the Arctic Ocean responds by creating a blanket of water vapor and clouds that keeps the Arctic warmer."

Christian responds to Dr. Pachauri and Professor White this way. "I understand UNO's IPCC has accepted that the rise of CO2 is the primary cause for the rise in temperature on Earth. And it is based on a model study by Penn State professor Mann and the hockey stick-shaped graph that shows an unprecedented sudden increase in average global temperatures. Per an article in Eurasia Review of November 12, 2012: 'Don't graphs show that current temperatures are the highest in 1,000 years?', the latest research clearly reveals that the Medieval Warm Period has been verified and was in fact global, not just confined to the Northern Hemisphere. And it states that 'the Medieval Warm Period was: (1) global in extent, (2) at least as warm as, but likely even warmer than, the current warm period, and (3) of a duration significantly longer than that of the current warm period to date. The Science and Public Policy Institute, Arlington, Va. reported that more than 700 scientists from 400 institutions in 40 countries have contributed peer-reviewed papers providing evidence that the Medieval Warm Period (MWP) was real, global, and warmer than the present.'

I reviewed the summaries and graphs provided in these papers. I think, it is reasonably accurate to conclude that the MWP or little hot age (LHA) had occurred in ~1,300AD (high point), with a temperature of ~1-2C warmer than the present time; and a little ice age (LIA) had occurred in ~1,800AD (low point). The evidence of tree-ring widths in ancient tree stumps buried under ground and discovered later show (i) larger growth with wider spaces in between the rings during LHA, and (ii) smaller growth with narrower spaces in between the rings during LIA. That gives a time period of ~500 years between LHA and LIA.

Using that as a guide we can build a calendar as follows:

Jesus (-30AD)*	~250 years before LHA-1(in 300AD)
Arya Bhata (500AD)	~200 years after LHA-1
Bhaskara I (650AD)	~350 years after LHA-1
Bhaskara II (1150AD)	~350 years after LIA-1(in 800AD)
Copernicus (1500AD)	~200 years after LHA-2(in 1300AD)
Kepler/Galileo (1600AD)	~300 years after LHA-2
Newton (1700AD)	~400 years after LHA-2 or
	~100 years before LIA-2(in 1800AD)
At present (2013AD)	~250 years before LHA-3(in 2300AD)

*Veda teacher 2,000 year ago

Per this scenario the present cycle began at LIA-2 about 250 years ago and it has to go another ~250 years before hitting a 'peak' temperature at LHA-3 in ~2300AD—this calendar is based on 1,000 year cycles per my calculations shown in Chapter 9. From the calendar above it is clear no major discovery or history came about during LHA or LIA, a time period of ~300 years each—Jesus words from Veda came in -30AD and the Romans adopted his 'words' in ~325AD. Does that mean Man has to wait for another ~300 years or until 2325AD to get the 'word' of Veda re-discovered here?

During the time of last LHA-2 there was no industry to create CO_2 and the humans were in millions, not in billions like at the present time—yet, it was warmer than the present time. So, I agree CO_2 or the man is not the primary cause for the 'rise in temperature on the earth'—it is the nature and my findings can prove that. The temperature on Earth gets hotter for about 250 years in a 1,000 year cycle. The temperature was hotter during LHA-2 ~800 years ago, with only few million humans on the earth, no industry, and CO_2 (at 400 ppm per Prof. White, University of Colorado). Now it is getting warmer (not as warmer as in LHA-2) with billions of humans on the earth, with industry, and CO_2 (at 400 ppm). The presence of excessive CO_2 in the air and breathing it could destroy some species on the earth—but, it is not the primary cause for the rise in temperature on the earth. It is so, because the CO_2 atoms in space do not crash into each other at high speeds or at the speed of light to split and generate the heat. Such an activity of fission is taking place in the sun generating the heat or radiation which disperses into the space. Naturally, when the sun and the companion-star, the two heat producing machines, come closer to each other the space in between becomes warmer; and when they move away from each other the space in between cooler. Therefore, the 1,000 year cycle in putting the sun and its companion on the same side of *Indra*, and the ever increasing spin of Earth's axis of rotation towards the sun during the next 720 cycles in this *yuga* is the primary cause for the rise in temperature on Earth or the coming Hot Age (HA). Therefore, it is the nature, not the man—the primary cause for 95% of the rise in temperature on the earth. The man is responsible for making 'garbage' and a part of the carbon-dioxide (CO_2)—detrimental to the life of the species. That is the reality—anything else could amount to a religion of the confused Man.

The pole reversal of *bhoomi* incorporates the *yuga* cycle of 4,320,000 years. A *yuga* is divided into four phases of unequal time periods. In the present *yuga*, we are in the last quarter of the third phase, with over 3,600,000 years passed. The survivors in the polar and equator regions from the Ice Age (IA) of more than 1,000,000 years have begun to multiply, and this outgrowth began about 30,000 years ago. Over the course of the remaining 720,000 years in this *yuga*, the orbiting sun and its companion-star would come onto the same side and go onto the opposite side of *Indra* once in a ~1,000 year cycle and will continue going for a total of 720 cycles before the end of time. During each cycle the living beings on Earth will face:

(1) Little hot age (LHA) with rising temperature for a quarter of the time (like at the present) as the sun moves closer to the companion-star on the same side of *Indra*—sun going in an inner orbit (elliptical) and companion going in an outer orbit (elliptical),
(2) A transition from hot to warm age for a quarter of the time as the sun moves away from the companion,
(3) Little ice age (LIA) with falling temperature for a quarter of the time as the sun moves farther away from the companion to the opposite side of *Indra*, and

(4) A transition from ice to warm age for a quarter of the time as the sun moves towards the companion—now, we are at the end of this period and back in (1).

The above four events continue to repeat in ~1,000-year cycles—like the earth going around the sun in elliptical orbit in ~365-day cycles changing seasons in four quarters. As a general rule in each 1,000 year cycle:

(i) the intensity in LHA should rise and the intensity in LIA should fall in the polar region facing the sun; and
(ii) the intensity in LHA should fall and the intensity in LIA should rise in the equator region and in the polar region which is not facing the sun—due to increase in tilt or spin of Earth's axis of rotation in each cycle. In reality, the actual intensity of heat in LHA or cold in LIA would vary in each cycle as a result of-

- a change in the actual distance between the sun and the companion in each cycle when they arrive on the same or opposite side of *Indra*—due to the orbital paths they follow, and a change in the sunspot intensity of the sun and the companion at the time.

The fourth and final phase of the present *yuga*, the end of time with Hot Age (HA), shall begin ~250,000 years from now—per the calculations given in Chapter 9.

I thank you Professor White for informing us of your findings. Your use of Pliocene period record of 3-5 million years old must be a mistake because (1) the record and the climate on Earth changes in 1,000 year cycles which is the missing link in the climate models, and (2) the 3-5 million years period is actually a 4.32 million year Cycle, and after each cycle Earth goes through a major change for tens of thousands of years with an end of life as we know it. So, the

man will never be able to find any clues about the CO2 of that time. And the rising temperature on Earth has nothing to do with rising CO2. The ice goes away in the hemisphere that faces the sun (which we call Arctic, now) in 1,000 year cycles or in the next ~200 years of this cycle. It is wrong to assume that the water vapor and clouds keep the Arctic warmer."

Another voice from the audience quickly raises "I am Dr. Carter and my friend Dr. Soon is with me—we are the experts at the Nongovernmental International Panel on Climate Change (NIPCC). We oppose Dr. Pachauri, IPCC, Al Gore, and the 97% of the scientists that claim 'man and his CO2 is the primary cause for the Climate Change'. We state they cannot prove that claim because there is no evidence—at NIPCC our only goal is to point out that. We, also, deny that the temperature on the earth is rising at all."

"I am Dr. Soon and I agree with my friend Dr. Carter that our purpose is to oppose IPCC. There is no evidence that the temperature on the earth is rising at all; or if the temperature is going up and down, it must be due to the fluctuating radiation of the sun and/or varying orbital distances of Earth in its wild rides like a drunk."

Christian thanks Dr. Carter and Dr. Soon for the comments and says "I agree CO2 is not, but Nature is the cause for the rise or fall of temperature on the earth in 1,000 year cycles due to the fact our-sun and its companion coming closer or going farther away. Simply opposing IPCC and 97% of the scientists may not help your cause until you can develop a solid evidence to show Nature is the cause. The suggestion of fluctuations in the radiation of the sun (or the companion) plus Earth's tilt is valid to explain the reason for the daily or monthly or yearly variations in the temperature on the earth. However, the suggestion that the orbital distances of Earth varies due to its wild rides is not helpful, without more, to prove the point—it appears all 100% of the scientists are confused or lost."

Another voice says "I am Dr. MacCracken, chief scientist at the Climate Institute, Washington, DC. I think CO_2 is the cause for the rise in temperature. However, it doesn't matter what is the cause if we can find a way to protect the species on the earth. The best thing to do would be to install a cloud cover in the arctic and see if it helps."

Christian responds "A cloud cover on the earth would act similar to a fur-coat worn by an Eskimo in the arctic cold to keep the body heat from radiating out and ice-cold from coming in. If the same Eskimo with a fur-coat would come into a hot region on the earth and refuses to take off the coat what happens—he will soon die. Likewise, if we place a cloud cover over the arctic (when it is hot), the species die."

Einstein asks, "Does the speed of light change in space?"

Christian answers, "Yes. The speed of light primarily depends upon the radiation, K, at a point."

Einstein asks, "How does the speed of light, $c=3x10^8m/s$, relate to K?"

Christian says, "The temperature of the radiation in space near Earth varies between 280–295K at the present time; thus the measured speed of light relates to 280–295K."

Einstein asks, "What would the speed of light coming from a distant star be?"

Christian says, "The speed of light varies on its way to Earth. And it shows up at about $3x10^8m/s$ when measured on Earth at 280–295K.

As the temperature on Earth increases in the hot climate the speed of light would increase."

Einstein asks, "Does that mean the astronomy as we know it is wrong?"

Christian affirms with a "yes."

Einstein says, "Explain why."

Christian says, "The present knowledge of astronomy is deduced based on observing only the visible matter, which is approximately 4 percent of the total matter. We know that the other roughly 96 percent of matter is invisible. Therefore astronomy based on that 4 percent must be wrong."

Einstein asks, "Do you know why the planets have elliptical orbits and orbit inclinations?"

"Yes," Christian confirms. "The *Indra* X-force causes elliptical orbits, and the *Indra* Z-force causes orbit inclinations to the planets."

The chair asks, "Have you discovered why wind speeds differ on the planets? For example, why does the wind speed increase as the distance from the sun increases? NASA observed that the wind speed on Saturn is greater than on Jupiter, which is greater than on Earth."

Christian confirms this with a "yes". "*Indra* is the cause for the increased wind speeds. We deduced an equation to determine the wind speed on a planet's surface. The equation gives a wind speed of 0 m/s on Mercury, being closer to the sun. The speed increases as the distance from the sun increases. *Indra* is the cause of the cyclones or tornados on the land, in the oceans, and on the sun."

The chair asks, "What are those five arms in Figure 8.3?"

Christian says, "They are the force bands between any two objects in space. The cold magnetic force 'plus' and the hot electric force 'minus' get hooked up to form bands and act as a tie or compression spring between *Indra*-sun, sun-earth, *Indra*-earth, earth-moon, and *Indra*-moon, like the DNA spiral helix that exists within *Indra* and keeps the sun, the earth, and the moon in their proper motions.

That brings attention to the ancient Veda symbols. We identify six key symbols passed on to us by an ancient *rishi* that confirms the accuracy of this work:

First, the sound of vibration, *Om*, created by *Sankhu*, sounds like the sound of vibration caused by GRBs within the Almighty *Brahman* and the living beings.

Second, the dark-*Swastika* with legs pointing to the left depicts the direction in which *Indra* rotates and the cold-*paramanuvu* orbits; the red-*Swastika* with legs pointing to the left depicts the direction in which Sun rotates and hot-*paramanuvu* orbits.

Third, the *Japamala* with 108 beads confirms that the length of the force-band between *Indra* and Sun equals to 108 AU, Astronomical Unit is the length of the force-band between Sun and Earth.

Fourth, the *Gadha* with large and small spheres joined by a variable diameter band, looks like *Indra* and Sun joined with a variable diameter force-band.

Fifth, the *Chakra*, an orbiting ring of fire, looks like Earth orbiting in Sun's fire around Sun's axis of rotation- ZZ-axis.

Sixth, the *Lotus* flower with petals looks like the Earth (or Sun) with magnetic-field flows from magnetic-N to the equator and equator to the magnetic-S."

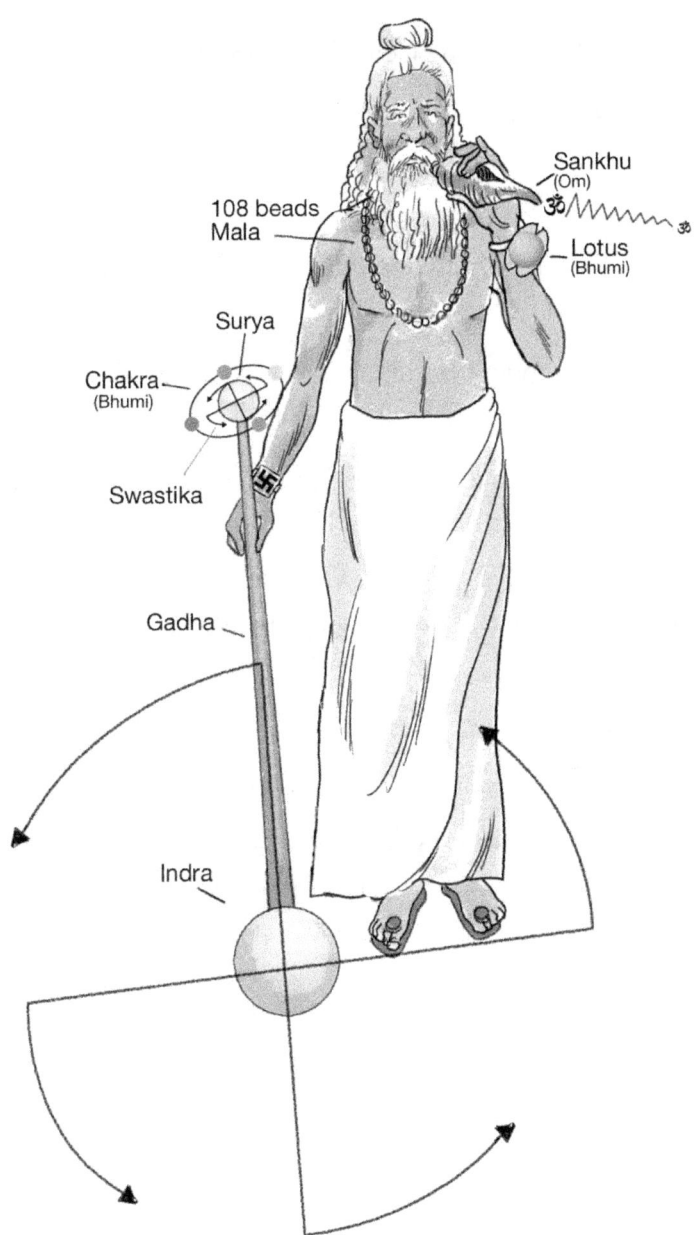

Fig.8.8: The Six *Veda* Symbols

171

Newton asks, "What is a galaxy?"

Christian says, "It is a living Cell of the *param-brahman*."

"I am Dawkins and an expert on god," says a voice. "How big is your god?"

Christian says, "The *param-brahman* is as big or small as you can imagine. This is the god of all the living beings on Earth and elsewhere."

Dawkins says, "I thought Brahman is a learned person in a Hindu temple."

Christian explains, "That is the Brahmin, the learned person in the Veda. In fact, you are a Brahman, a little Brahman, to the creature living within you."

Dawkins asks, "You think the god is made of Atoms and Cells like we are?"

Christian says, "Yes, all living beings are created in God's own image, an Atom."

"I am Darwin," says another speaker, "of the family of the real Charles Darwin. I am a true believer in Darwinism, or evolution, like all the scientists on this earth. If 'god' is made of atoms, like living beings, your work must affirm Darwinism—that all living beings are evolved from an atom. So I ask if you agree that Darwin was correct."

Christian answers, "No, I do not. If all living beings evolved from the same atom, we all should look alike, including the fish and ape. They do not. Therefore Darwin's concept of evolution is nonsense."

Dawkins says, "So can you see god within you?"

Christian answers, "No, I cannot see god directly. I can see the presence of god in all the living beings and everywhere."

Dawkins says, "Please tell me how you see this."

Christian says, "Consciousness is god, because it senses the light, sound, smell, taste, and touch."

Dawkins says, "Are you telling me god is within me because I sense all that?"

Christian answers, "Yes."

Dawkins says, "Please explain."

Christian answers, "You are made of atoms and infinitesimal atoms in a variety of shapes and sizes. Each organ in your body is made of different types of atoms and infinitesimal atoms, and each one group of these atoms performs a specific duty to get you going. As an example take your eye—the receptors receive light and transmit it into your brain to enable you to see or that which makes you see is your consciousness. Only the living infinitesimal atoms can do that, not the dead or unconscious ones. So consciousness is the act of 'god.'"

The chair asks, "Did you publish this work in a science journal?"

Christian says, "No, we are going to publish it in a book."

The chair asks, "Why publish a book and not in a science journal?"

Christian says, "If there was a choice we would have published it in a science journal and a book. But we had no choice. The journals did not want to publish anything connected to or about 'god.' The editors and scientists are satisfied with god-Darwin and god-Einstein and believe there is no other god. Therefore they refused to understand this work about Reality. Also, they think this work doesn't serve any purpose because the curvature of space-time (GR) is the reality."

The chair says, "Tell me what makes the signal move in a computer chip or wireless phone or the Internet, and what makes an asteroid orbit in space?"

Christian says, "It is the *paramanuvu* in space. The *shakti* in the form of entropy or the change in temperature that man imparts into the atom makes the signal move. In the case of an asteroid, it is like a mini-earth upon which the *Indra*-sun imparts that entropy to make it move."

The chair says, "One final question. Why so much confusion about what is Reality?"

Christian says, "The *karma* of our time."

The chair says, "Thank you all. That concludes this part of the presentation. Well done. Those of you wish to listen to the reading of the research paper, please stay on to understand the details, such as the equations and calculations, etc."

Do we survive the hot climate, we do
Do we succeed in the adventure, we do
Do we succeed in the detective work, we do
Do we discover Veda, we do
Do we learn Veda, we do
Do we learn *Isvar*, we do
Do we learn *Isvar* controls god-*Indra*, we do
Do we learn god-*Indra* controls *Surya-Bhoomi* and *Graha*s, we do
Do we learn *Indra*, *Surya*s-*Graha*s make one-*Anuvu* of *Brahman*, we do
Do we learn living beings on *Bhoomi* live within the *Brahman*, we do

Do we learn *Isvar* is within the *Brahman* and everywhere, we do
Do we learn *Isvara* is within the living beings, we do
Do we know the time is ripe for Pharisee and soldiers to learn Veda, we do.

aatmaanam rathinam viddhi

Sareeram rathamevatu

Buddhim tu saaradhim viddhi

Manaha pragrahamevacha

Know the Self as the Lord of the Chariot and the body as the Chariot itself
Know the intellect as the charioteer and the mind as the rein
(*Kathopanishad*)

Fig.8.9: The three Professors sing and dance…

Glossary:

Maya= Mystery
*Indra-loka=*Space of *Indra*

Chapter: 9

The Research

Christian says, "Now in this session, I am going to present the research paper in its entirety with some help from Professors Esther and Mohammad. We can discuss your questions, if any, at a later time, after you had a chance to review the material carefully in your own time. Therefore, at this time, please allow me to read the paper": The structure of our-universe and the motion of matter- Part 1" by *Maheswar*

Summary:

harih aum, s'am no indro, namo brahmane,
namaste vaayo, ritam vadisyaami, satyam vadisyaami,
aum santih, santih, santih!

Om, may *Indra* be propitious to us
I will speak of the right, speak of the truth
Om peace, peace, peace!
(*Taittiriya Upanishad*)

In this work, we deduced the 'structure of our-universe' with an *Indra* (Neutron) at the center, not the *Surya* (Sun). *Indra* force acts upon all the objects in the *Anuvu* (Atom). *Indra* makes *Surya* orbit around *Indra*, and rotate. *Surya-Indra*, together, make Planets orbit around *Surya* at an orbit-inclination, and rotate with wobble and spin of its axis of rotation which causes variable gravity and dynamic weather on the Planet's surface. The leading point in its orbit, known as aphelion, moves away from *Surya* and closer to *Indra*, and the trailing point in its orbit, known as perihelion, moves closer to *Surya* and away from *Indra* which results in an orbit inclination to the plane of *Surya-Bhoomi* (Earth) line or elliptic. The *Anuvu* is a fluid ball, an elongated-sphere of variable diameter ~722 AU, with submerged objects: 2-*Surya*(s), Planets and Moons etc. that move within and around *Indra*- like in a live Helium atom-He. The observed (~2.7K) cosmic microwave background (CMB) is the left-over radiation of *Surya* at *Indra*. Change in the red shift is local phenomena within an *Anuvu*. The temperature, T, and entropy gradient directly affects change in the red shift and the speed of light. The speed of light, c, on *Bhoomi* relates to T ~280-295K.

*Sanskrit in *Italic*

<u>Text</u> -Background:

trinaabhi chakramajaramanarvam yenemaa |
visvaa bhuvanaa nitasttuhu ||

The path through which the planets are moving…
That path is eternal, indestructible and a tri-centered circle or elliptical in shape.
(*Rig Veda*)

Kepler law (1601CE) gives a description of the motion or orbit of a Planet around the Sun as an ellipse based on Mars orbit-data observed by Brahe. Kepler did <u>not</u> know <u>why</u> a lone Planet (of near zero-mass) moves around the Sun this way. A century later, Newton showed (1687AD) that Kepler law is a natural consequence of his inverse square law of gravity force, and deduced the forces that keep the Planet in its orbit, which must be reciprocally as the square of its distance from the center about which it revolves.

Newton did not know why the Planets move at various distances from the Sun; nor did he deduce the cause of the properties of gravity or of the force transfer between Sun-Earth. He derived an equation for a circular-orbit of radius, R, an average-distance between Sun-Planet. More than two-centuries later, [1&1.1]Einstein explained that the gravitational force is a fictitious force due to *space-time curvature; but, he could not answer why the varied orbit-distances of the Planets from the Sun or why the varied orbit inclinations to the elliptic. So, the premise of his theory is fictional, not the reality.

*Einstein's 'The Foundation of General Theory of Relativity', DOC.30, says "For infinitely small four-dimensional regions the theory of relativity in the restricted sense is approximate, if the coordinates are suitably chosen. For this purpose we must choose the acceleration of the infinitely small (local) system of coordinates so that no gravitational field occurs; this is possible for an infinitely small region" see @4, p.154
"Our next task is to find the field Eqs. of gravitation in the absence of matter" see @14
"..gravitational fields which may be regarded as generated exclusively by matter in the finite region" see @21

In essence, Kepler showed us a concept to derive an equation of an ellipse describing an orbit in a two-body problem, using the data on Mars orbit around the Sun. Kepler had no clue what is causing such a radical, non-circular, orbit of a Planet around the Sun. Then, Newton showed us a concept to derive an equation describing gravity in a two-body problem involving Sun-Earth. Later, others have come up with methods for solving an n-body problem representing the objects that orbit the Sun; but none successfully. Even if successful, a solution to an n-body problem can only describe the status, as Kepler or Newton did, but it would not disclose the cause for the varied orbit-distances from the Sun or the varied orbit inclinations to the elliptic, including the retrograde orbits of the outer Moons.

In May 2011, National Aeronautical Space Administration (NASA) presented results on its gravity probe-B experiment in space, and asserted that the 'geodetic plus frame dragging effect of space' near Earth, per "[1]Einstein's theory of general relativity or space-time curvature, is a drift of ~6.6 arcsec/year in reference to a Star. Per NASA Einstein's universe is curved* and Newton's is flat; even if it is true, NASA could not explain <u>how</u> this drift relates to the varied orbit-distances of the Planets from the Sun, or <u>why</u> the varied orbit inclinations of the Planets (or Moons) to the elliptic?

The answer to the mystery lies in the reality of the 'structure of our-universe' that verifies the observed NASA data. There can<u>not</u> be any curved-space or *deflected-light ray, other than hot-matter moving away from Sun surface (corona) at high-speeds in spiral-circles, and an absence of light during eclipse in the area covered by Moon-shade (which causes diffraction of the light, a Raman affect). The premise of Eddington experiment (1919CE) in measuring deflected-light ray of a star by comparing two-photos, one taken during Moon-Sun eclipse and another a day or a month later, is erroneous- because no one can determine an exact position of the background star(s) looking at such photos as the stars would <u>not</u> stay put, but keep moving. A slow motion in the orbit of *Mercury at 0.43arcsec/year must be an error in the reality as the orbit-speed of Mercury varies, not constant.

*Einstein in 'The Foundation of General Theory of Relativity', DOC.30, says "The ray of light going past the sun undergoes a deflection of 1.7" (arcsec); and the ray going past planet Jupiter a deflection of about 0.02" (arcsec). The orbital ellipse of a planet undergoes a slow rotation in the direction of motion per revolution [Mercury: 43" (arcsec) per 100 years]" see @22

References on the observed <u>Mystery</u> *v.* <u>Cause</u>:

[1] Doyle et.al., *"Kepler-16: A Transiting Circumbinary Planet"* (Science, Vol. 333 no.6049 pp.1602-6, published on September 16, 2011; NASA Website) report: **"detection of a planet whose orbit surrounds a pair of low-mass stars. (Photometric) data…reveal transits of the planet across both stars, in addition to the mutual eclipses of the stars…(and) the planet is comparable to Saturn in mass and size… Eclipses occur because the orbital plane of the stars is oriented nearly edge-on as viewed from Earth… During primary eclipses (larger) star A is partially eclipsed by the (smaller) star B, and the system flux declined by ~13%; during secondary eclipses B is completely occulted by A and the resulting drop in flux is only ~1.6%... Three additional drops in brightness were detected outside of the primary and secondary eclipses, separated by intervals of 230.3 and 221.5 days. These tertiary eclipses could not be attributed to the stars alone, and indicated the presence of a third body. The differing intervals between the tertiary eclipses are simply explained if the third body is in a circumbinary orbit, because stars A and B would be in different positions in their mutual orbit each time the third body moved in front of them. In contrast, there would be no ready explanation for the shifting times of the tertiary eclipses if they were produced by a background star system or some other unrelated event. During tertiary eclipses, the total light decline of 1.7% > 1.6% .. (so it) had to be transits of the third body across star A. This interpretation was supported by the subsequent weaker 0.1% quaternary eclipses which were consistent with the passage of the third body across star B (and two other quaternary eclipses were detected)… (The) third body was shown to be a transiting circumbinary planet (in the) model based on the premise that the three bodies move under the influence of mutual Newtonian gravitational forces."**

NASA's Website says- "Given that most stars in our galaxy are part of a <u>binary</u> system, this means the opportunities for life are much broader than if planets form only around single stars… Most of what we know about sizes of stars comes from eclipsing binary systems, and most of what we know about the sizes of planets comes from transits. Kepler-16 combines with stellar eclipses and planetary transits in one system."

<u>Cause</u>: First, eclipses would not occur if the orbital plane of the stars is oriented nearly edge-on as viewed from Earth; eclipses occur when Planet or Moon comes into a line of sight with a Star (or Sun) having the orbital plane of the planets oriented nearly edge-on to our (Kepler) line of sight. Second, eclipses would not occur because star A eclipses star B or vice versa; even if that occurs the system flux would not decline, but increases. The data collected in ~600 days show (Doyle Fig.1) a total ~15 (primary-eclipses), ~15 (secondary eclipses), ~3 (tertiary eclipses) and ~3 (quaternary eclipses). If we place Kepler at ~10,000AU away from Earth and collect similar data for ~686 days looking toward our-Sun, we will see eclipses!

So, counting them with a first-eclipse of Sun-Earth:
Sun-Earth ~1+(686/365)=2.88, Sun-Venus ~1.88/0.615=3, Sun-Mercury ~1.88/0.241=8 & Sun-Mars ~1.88/1.88=1: Total 15 eclipses at our-Sun. Since the 2-Suns orbit Neutron (in our-Atom) at different speeds and radii, the companion-Sun will be closer to Kepler, being the outer-Sun. So, the eclipses at the companion-Sun would be primary, and the eclipses at our-Sun would be secondary with a more decline in the flux compared to the eclipses at the companion-Sun. Additionally, Kepler would see eclipses from companion Sun-Moon ~**3** tertiary (equal to companion Sun-Earth ~3), and from our Sun-Moon ~**3** quaternary (equal to Sun-Earth ~2.88). Kepler would see

more eclipses attributable to Jupiter, Saturn, Uranus & Neptune etc... should the collection of data continued beyond 686 days.

From the data provided by Doyle, we can find a similarity between the binary stars A-B and our Atom, in which our Earth orbits. The only difference being in the AU distances between Sun-Planets. There, Kepler could see ~15 primary and ~15 secondary eclipses in 600 days, and here it takes ~686 days to see the same number of eclipses. Despite the similarity in the 'structure' of these two-Atoms, the difference in AU distances alone would rule out any possibility of life there, similar to what we have here on the Earth. There, the primary or secondary eclipses cannot be due to transits of star A or B, because if it were due to the mutual eclipses of stars the relative flux must increase, not decline ~13% or ~1.6%. Like here, the orbital plane of the planets there, must be oriented nearly edge-on to view.

Single star-planet system cannot exist since it is unstable; and, the sizes or masses of stars (in a galaxy) cannot vary since the system become unstable. A Planet cannot transit across two-stars under 'Newtonian gravitational forces'. It is simply incorrect to model or explain that the third body, a Planet, is in a circumbinary orbit there. Thus, the finding of the eclipses between planets-star A and planets-star B confirms most (or all) stars in our-galaxy are binary systems, including ours.

[2] Burrows, et.al., "Relativistic jet activity from the tidal disruption of a star by a massive black hole" and Zauderer, et.al., "Birth of a relativistic outflow in the unusual y-ray transient Swift J1644+57" (Nature 476 p.421-4 & p.425-8, Aug. 25, 2011) report details how a distant black hole devoured a star. This is the latest on Einstein-Schwarzschild metric (BH) (see *[1a]NASA Website, August 24, 2011). Burrows reports that the event is probably due to the tidal disruption of a star into a black hole, estimating the total luminosity of the

source 4.2x10^46 erg/s and infers the luminosity relative to our-Sun equal to ~9.2-9.58; and that leads to a black hole (BH) mass of 2x10^7M or 7x10^6M (M=Suns' mass). The measured red-shift of 0.354 corresponds to a luminosity distance D of 5.8x10^27cm (using L=Fx4piD^2, where F is X-ray flux). Fig. 2 shows X-ray flux at ~5x10^-10 erg/cm^2/s. And, Zauderer, with EVLA at 5.8GHz and at 1-345GHz found a predicated SSC X-ray luminosity 2x10^45 erg/s and a modest BH mass of 10^6M.

Cause: There is no surprise here- in May 2011, using erroneous-results from gravity probe-B experiment in space, NASA asserted proof of most profound predictions of *1Einstein's universe; and in August 2011, it uses Burrows, Zauderer et.al to assert proof of *1aSchwarzschild's description of a BH, a brain-child of Einstein's theory of general relativity (GR). In reality, BH does not exist and GR is a fiction. The 2-articles estimate different BH mass and different luminosity for the ripped-star, and could not determine the flare-temperature, T. One of the named author agreed (on phone) that "it is a guess the BH ripped the star…".

Taking star's X-ray flux at ~5x10^-10erg/cm^2/s (600km above the Earth), and using Sun's X-ray flux at ~5x10^-4erg/cm^2/s from GOES15 (35,800km above the Earth), and the values from Table 4 (with Neutron existing)- we determined the actual distance between Earth-star, Da~1,060,000AU and star's flare-temperature, T ~60x10^6K.

[2.1] Perlmutter "Supernova, Dark Energy and Accelerating Universe" (see UCSB Website) reports: "In principle, the expansion history of the cosmos can be determined quite easily, using as a standard candle any distinguishable class of astronomical objects of known intrinsic brightness that can be identified over a wide distance range. As the light from such beacons travels to Earth through and

184

expanding Universe, the cosmic expansion stretches not only the distances between galaxy clusters, but also the very wavelengths of the photons enroute. By the time the light reaches us, the spectral wavelength, λ, has thus been red-shifted by precisely the same incremental factor $z=\Delta\lambda/\lambda$ by which the cosmos has been stretched in the time interval since the light left its source. That time interval is the speed of light times the object's distance from Earth, which can be determined by comparing its apparent brightness to a nearby standard of the same class of astrophysical objects. Conceptually, this scheme is a remarkably straightforward means to a profound prize: an empirical account of the growth of our Universe. In Hubble's discovery of the cosmic expansion (in 1920s), he used entire galaxies as standard candles... galaxies coming in many shapes and sizes are difficult to match against a standard brightness. ... The faintness or distance of the high-redshift supernovae was a dramatic surprise... the high-redshift supernovae are fainter than would be expected even for an empty cosmos... there remain the all-important questions of systematic uncertainties.. the two groups' efforts have been devoted to hunting down these systematics. Could the faintness of the supernovae be due to intervening dust? The color measurements that would show color-dependent dimming for most types of dust indicate that dust is not a major factor...By confirming the flat geometry of the cosmos, the recent measurements of the CMB have also contributed to confidence in the accelerating Universe results... one would have to invoke improbably large systematic error to negate the supernova results... (that) explodes somewhere in the sky every few seconds".

Perlmutter UCSB on "High Red-Shift Supernova" (see UCSB Website: Experimental Cosmology Group) reports: "A supernova is an exploding star... extremely bright and causes a burst of radiation that often briefly outshines an entire galaxy, then fades over several weeks… Supernova do not show strong hydrogen lines in its spectra, and arises from a binary system, the type 1a shows strong silicon absorption lines at 615 nm (6150 A), and makes a good standard candle- at high red-shifts, for determining cosmic distances and to measure the expansion history of the universe. Normally, the spectra peak in the blue-violet region in its own rest-frame; so we observe it in a red-infrared filter, because the blue-violet light is red-shifted. If we observe it in blue-violet filters, we won't measure their peak light output, in our rest-frame. The luminosity of the high-z supernova appears too faint for its red-shift, .. (and) provides independent evidence for the cosmological constant. The best fitting model indicates that there was an increase in the expansion rate of the universe at a red-shift close to $z=0.5$, corresponding to approximately 5 billion years ago, in the history of the Universe".

Knop et.al. "New Constraints… from an Independent set of Eleven High-redshift Supernovae Observed with HST" (see arXiv:0309368 v1, Sept. 12, 2003) report: ".. one obvious possible source of systematic uncertainty is the effect of host-galaxy dust… Dust extinction from within the host-galaxy… could have (made the observed brightness dimmer..); however, normal dust will also redden the colors of the supernovae. Therefore, a measurement of the color of the high-redshift supernovae, compared to the known colors of low-redshift supernovae, has been used to provide an upper limit on the effect of host-galaxy dust extinction, or a direct measurement of that extinction which may then be corrected. Uncertainties on extinction corrections based on these color measurements usually dominate the statistical error of photometric measurements… Each supernova is imaged with 2-broadband filter using CCD of the WFPC2 on the HST… (and) photometric fluxes are extracted from the final images…"

<u>Cause</u>: Perlmutter 'principle of expansion history of the cosmos based on supernovae results, which is based on Hubble discovery of the cosmic expansion' (1920sCE), **negate due to large systematic errors** as follows:

Let us see Kirshner (Hubble's diagram and cosmic expansion, PNAS January 6, 2004 Vol 101 no.1 8-13) report which said: "Hubble showed that galaxies recede from us in all directions and more distant ones recede more rapidly in proportion to their distance... Although there were hints of cosmic expansion in earlier work, this publication convinced the scientific community that we live in an expanding universe... Hubble was able to show...more astonishing by plotting the velocities of galaxies against their distances... The quantitative agreement of modern measurements with Hubble's original distance scale is not good! Modern distances to the same galaxies, reckoned to be accurate to 10%, are seven times larger than the distances Hubble plots... Astronomers measure the velocity of a galaxy from its spectrum by taking the light from a galaxy's image at the focus of a telescope and passing it through a slit and a prism to create a dispersed rainbow, subtly marked by dark lines... The fractional shift of the wavelength, $\Delta\lambda/\lambda$, is $1 + z$, where z is the redshift. This result can be expressed as a velocity $v=cz$ (where c is the speed of light)...

The measuring of galaxy spectra.. was initiated .. by Slipher at the Lowell Observatory in Arizona... (and) compiled a list of velocities for 41 galaxies (1923CE) of which 36 were receding from us (with a largest velocity of 1,800 km/s).

The galaxies constructed from Hubble's Law is surprisingly foamy, with great voids and walls that form as dark matter clusters in an expanding universe, shaping pits into which the ordinary matter drains, to form the luminous matter we see as stars in galaxies...

Hubble was very circumspect.. on the question of whether cosmic expansion revealed a genuine cosmic history. He referred to the redshift as giving an "apparent velocity." In a letter to Willem de Sitter, Hubble wrote, ".. We use the term 'apparent' velocities to emphasize the empirical features of the correlation. The interpretation, we feel, should be left to you and the very few others who are competent to discuss the matter with authority."... Einstein's idea of a static universe suspended between gravity pulling inward and the cosmological constant making the universe expand was ruled out by Hubble's data... Supernova Cosmology Project showed the surprising result that the expansion of the universe has been speeding up during the 5-billion-year interval while the light from a distant supernova has been in flight to our telescopes... As Hubble said "We measure shadows, and we search among ghostly errors of measurement for landmarks that are scarcely more substantial. The search will continue". Hubble's article ... was not enough to convince Hubble himself of the reality of cosmic expansion, but that article in PNAS pointed the way to understanding the history of the universe, and the continuing search among the "ghostly errors of measurement" has led to a deeply surprising synthesis of dark matter and dark energy..."

Thus, Hubble was '**not convinced himself of the reality of cosmic expansion**' because he knew the velocity, $v=cz$, is only an 'apparent' velocity- not the real to cause expansion of the Universe. Hubble knew or should have known of the work (1864-8CE) by a great man, William Huggins "Further Observations on the Spectra of some of the Stars...to determine there from whether these bodies are moving towards or from the Earth (see Philosophical Transactions of the Royal Society of London May 14, 1868, p.529). That work, a master piece, reports-

"If the velocity of light be taken at 185,000miles/s.., the observed alteration in the period of the line in Sirius will indicate a motion of recession existing between the Earth and star of 41.4 miles/s. Of this motion a part is due to the **Earth's motion in space...in the plane of elliptic,...changing the direction of its motion at every instant... it is moving in the direction of the visual ray...either towards or from the star... or at right angles to the...light from the star**" (p.548).

"At the time when the estimate of the amount of alternation of period of the line in Sirius was made, the Earth was moving from the star with a velocity of about 12 miles/s. There remains unaccounted for a motion of recession from the Earth amounting to 29.4 miles/s, which we appear to be entitled to attribute to Sirius. It is interesting, .. to refer to the remarkable inequalities which occur in the rather large **proper motion of that star**. In 1851CE, Peters showed that the **variable part of the proper motion of Sirius** in RA might be represented by supposing that **Sirius revolves in an elliptic orbit round some center of gravity** without itself, in a period of 50 Yrs.. and seems to have received confirmation from direct observations by Clark's discovery of a small **companion to Sirius**. ...Prof. Safford and Dr. Auwers have investigated the periodical **variations of the proper motion**..., and found.. (to) be reconcilable with an **elliptic orbital motion round a center not in Sirius**.... We must then suppose it to have a much greater mass relative to Sirius, than that which its light would indicate" (p.549).

"It may be that **in the case of Sirius we have two distinct motions, one peculiar to the star, and a second motion which it may share in common with a system of which it may form a part**" (p.550).

Huggins discovery (1868CE) made it clear that Sirius, too, moves in an elliptical orbit like the Earth. Given that, it is not hard to see that at certain period Sirius appears to be receding towards or from the Earth as they keep orbiting in their own plane of elliptic and at their own radial velocities… In fact, Slipher's evidence compiled (1923CE) lists velocities for 41 galaxies, of which 36 were receding from us- see Kirshner. The other 5 galaxies must be either receding towards us or moving at right angles to the direction of the visual ray, in which case Slipher could not observe or determine their apparent velocities… In other words, we (being on Earth) move in our own plane of elliptic (around our-Sun) while Sirius or a distant-galaxy moves in its own plane of elliptic (around and within a Neutron). Therefore, at certain period Sirius or a distant-galaxy appears to us as if it is receding from us and at other times appears as if it is receding towards us… That does not mean Sirius or the distant-galaxy is accelerating away from us causing expansion of the Universe. Although Einstein's idea of a static universe is 'correct', his concept of gravity pulling inward and the cosmological constant making the universe expand is 'incorrect'- because 'gravity' is the resultant force at a given point and varies from point to point in the Universe like entropy or red shift or speed of light.

Kirshner, also, said "the galaxies constructed from Hubble's Law is surprisingly foamy, with great voids and walls that form as dark matter clusters in an expanding universe, shaping pits into which the ordinary matter drains, to form the luminous matter we see as stars in galaxies". Here, the so called 'foamy great void or the dark matter cluster' must be the Neutron (*Indra*) that controls the galaxy, like a Neutron controlling binary stars, such as ours, Sirius or the one in Doyle[1]; obviously, a galaxy, too, must orbit around a Neutron like the stars do. The apparent velocity of a galaxy makes it appear as if it is an 'expanding universe' like the mirage in a dessert. Neutron is the cause for the 'matter drain to form as stars'. Whether 'astronomers measuring the velocity of a galaxy from its spectrum by taking the light from a galaxy's image at the focus of a telescope and passing it through a slit and a prism to create a dispersed rainbow, subtly marked by dark lines' is correct [see Part 2 & 3]. Suffice to state, here, the atoms in the prism material create the dispersed rainbow of colors (although the light-ray has no color), and the atom neutrons create the 'dark lines in the rainbow'; and the supernova distance and temperature can be determined from the photometric flux (CCD of the WFPC2 on the HST: see Knop) as shown at [2]. The "ordinary matter drains to form… as stars in galaxies" or mysterious cosmic 'dark flow'[4] can be related to the 'blood-cells' draining into the lungs, liver... to be purified with fission, a process similar to the "supernova exploding somewhere in the sky every few seconds" (*Om*).

And, the "host-galaxy dust (and) reddening the colors of the supernovae" is similar to the reddening we observe at sunrise or sunset- which occurs when Sun-ray or supernova-ray directs toward the Earth at a particular angle through the hot-*params* in space. Thus, the dust and redden-color of supernova defines the angle between the elliptic planes of the Earth and supernova…

[3] Dark fireworks on the Sun (see NASA Website).
In June 2011, NASA's SDO detected a blast of filament on Sun's surface; the filament blobs (as big as a planet or larger than Earth) loaded with cool plasma (~20,000K) and gas (~1×10^6K) rose like balls tossed in the air, and moving under the gravity of the Sun fell on the stellar surface. "The sunspot above the one associated with the blast attracted the ejected plasma, appearing much like a smoker inhaling. The blast propelled coronal mass injection (CME) out of the Sun's atmosphere. The amount of material that fell back to the Sun is about equal to the amount that flew away".

Cause: Dark fireworks from a sunspot blast of ejected gas ($\sim 1 \times 10^6 K$) rising, propelling coronal mass injection (CME) out of the Sun's atmosphere, and cool plasma ($\sim 20,000K$) falling into another sunspot above, under the gravity of the Sun is like the living exhaling hot-air and inhaling cold-air in equal amount. Neutron, the maker of cold, is the cause in the living.

[4] Mysterious Cosmic 'dark flow' (see NASA website: Study published in ApJ letters on 03/20/2010): Kashlinsky of NASA (GSFC) "tracks the collective motion of 'dark flows',.. The clusters appear to be moving along a line extending from our Solar system…, but the direction of this motion is less certain. Evidence indicates that the clusters are headed outward…, away from Earth, but… cannot yet rule out the opposite flow… The dark flow is controversial because the distribution of matter in the observed universe cannot account for it. It.. suggests that some structure beyond the visible universe- outside our horizon- is pulling on matter in our vicinity."

Cause: The find of mysterious cosmic 'dark flow' away from Earth or towards Earth, which is "pulling on matter in our vicinity (by) some structure beyond the visible universe", must be the Neutron; and the flow must be of the Neutrons in a cluster or clusters…

[5] N. Lehner et.al., 'A Reservoir of Ionized Gas in the Galactic Halo to Sustain Star Formation in the Milky Way' (*Science* DOI: 10.1126/ science. 1209069, Published Online August 25 2011) says: "We show ionized high-velocity clouds (iHVC).. are within one galactic radius of the Sun and have enough mass to maintain Star formation.." (in our-galaxy) with "cold stream accretion as a means for metal-poor gas to flow onto galaxy along dense intergalactic filaments.. galaxy may also exchange with the .. outflows driven by galactic feedback.."

<u>Cause</u>: Galaxy must be a living-Cell that operates under the cold-force of Neutrons, within and without, as the 'dark flow'.

[6] CERN's ALPHA Collaboration (see Nature Physics 7, 558-64, June 05, 2011) Confined anti-hydrogen for 1,000 seconds annihilated by hydrogen;

[6.1] Sun gives the radiation (heat)- who gives the cold (ice)?

<u>Cause</u>: CERN's anti-hydrogen is the cold matter (per CERN); and there is no-source other than Neutron in our-universe that can give the cold, or make the cold-dark matter orbit. The mystery of matter *v.* anti-matter is no more than hot-matter converting back into cold-matter giving-up its radiation, and in the process nothing annihilates (Caveman knew that to be a fact).

[7] S. M. Krimigis finds (from the data of NASA Voyager 1 & 2, see Nature 474, 359-61, June 16, 2011), Zero outward flow velocity[*2, 2.1, 2.2 & 2.3] of plasma at a distance ~116 AU from the Sun.

<u>Cause</u>: Krimigis finding that the flow velocity of ~100 km/s at a distance ~100AU from our-Sun confirms the result obtained in this work. See Figures 1-2, <u>Tables</u> 2-6 for the deduced 'structure of our-universe' and 'motion of matter'. However, Krimigis and rest of the scientists including NASA do not have an accurate answer to the issues, such as-

(i) What makes the Sun orbit or the direction of orbit, and its speed?

(ii) Where are the Voyagers (1 & 2)? Did they reach 'heliosheath' (nose) or the 'tail end' of the 'heliosphere'?

(iii) What is the cause of the 'ribbon-knots' observed by IBEX, and where is its location...?

This work provides answers to these and more...see[*2.1-2.3]

[8] Saturn and Jupiter rings corrugated (looking along the equatorial plane: see NASA Website).

Cause: The corrugations of the ring-objects at Jupiter, Saturn is no different from the corrugations of the Planets at Sun where at perihelion or aphelion Planets go up or down as the orbit inclinations (Pi) of each Planet vary… under the direct-force of Neutron. That we can see if we join lines from Sun-Mercury, Mercury-Venus, Venus-Earth…and so on. The same direct-force of Neutron causes the corrugations of the ring-objects at Planet Jupiter and Saturn because the ring-objects, too, move up or down like the Planets, with varied orbit inclinations.

[9] D.H.Hathaway (MSFC), 'The Solar Cycle'[*2a] [Living Rev. Solar Phys. 7, (2010CE), citing H.W.Babcock, 'The Topology of the Sun's Magnetic Field and the 22-year Cycle' (ApJ 133, 572-587)] reports: "Sunspots erupt in low latitude bands on either side of the equator and drift toward the equator as each cycle progresses. Fig.8: shows two bands, one in each hemisphere, at about 25° from the equator. …the sunspot-groups extend in longitude, more constrained in latitude, with one magnetic polarity associated with the leading spots (in the direction of rotation) and the opposite polarity associated with the following spots. … The magnetic polarities of active regions reverse from northern to southern hemispheres and from one cycle to the next. The polar fields reverse polarity during each cycle at about the time of cycle maximum… Diffusion of the erupting active region magnetic field (at the equator) then transport by the meridional flow leads to the accumulation of opposite polarity fields at the poles and the ultimate reversal of the polar fields... Fig.14 (a magnetic butterfly diagram constructed from the longitudinally averaged radial magnetic field obtained from instruments on Kitt Peak and SOHO) illustrates.. polar field reversals, and the transport of higher latitude magnetic field elements toward the poles".

Cause: "…the transport of higher latitude magnetic field elements toward the poles" cannot be true; the analysis proves that as shown in Figures 1b & 1c. In a given cycle, if we assume Sun's magnetic-N is pointing toward the Neutron, then "the transport…toward the poles" can be true as some elements from the S-hemisphere move toward magnetic-N under Neutron's gravity while some elements move toward the equator to diffuse; and the elements from the N-hemisphere transport toward magnetic-S without much effect from Neutron's gravity. A similar transport or flow of the elements must be true in the case of planets under Sun-Neutron gravity, and for the Moons under Planet-Neutron gravity.

[10] J.Chadwick, 'The existence of a Neutron' (Proc. Roy. Soc., A, 136, p.692-708(1932)): His hypothesis based on indirect evidence (p.700-2) on the 'existence of a neutron' in Hydrogen (H) says "neutron…consists of a proton and an electron in close combination…, (and) find…the mass of the neutron is 1.0067x (mass of proton + electron)"- an invisible-insensible particle.

Cause: Physicists readily accepted Chadwick's hypothesis without seeing or sensing the neutron; and no one has any clue as to what this 'neutron' does in an atom- other than to assume- it is the 'keeper of proton in the atom-nuclei', and make 'electron' the actor. Like Chadwick's simple equations and analysis- "on radiation, the effects of which have been examined…, consists of neutral particles"- here, we present simple equations and analysis that shows the 'existence of a Neutron' in our-universe (Atom) which verifies the observed NASA data in space. Only difference being- Chadwick's hydrogen consists of 1 x mass of (proton + electron), whereas, our live Atom consists of 2 x mass of (Sun + planets). The 1 x mass of (Sun + planets) would not verify the observed NASA data. Astronomers agree we see double-Suns in the Sky (in our neighborhood). In alpha decay (Curie

1903CE), a natural process of decay, matter on the Earth loses in the form of He (2 neutrons + 2 protons +2 electrons); but, never observed losing in the form of H (1 neutron + 1 proton + 1 electron).

NASA finds (see its website):

[11] Radical, not circular, orbits of Planets at varied orbit-distances from the Sun (that made Kepler, Newton wonder…),

[12] Varied orbit inclinations to the elliptic, with greater inclinations of the outer Planets (that no one could solve why…),

[13] Varied orbit inclinations of the Moons of Jupiter, Saturn, and Neptune including retrograde (or polar) orbits of outer Moons while the inner Moons orbit in the direction of Planet's rotation (that no one has a clue why…),

[14] Wind-speed on a Planet increases as its distance from the Sun increases (per NASA new finding wind speed on Saturn>Jupiter>Earth). It defeats all logic <u>why</u> the wind speed on Mercury equals to ~0 (being close to the Sun); what causes cyclone or tornado (in Ocean, on Earth or Sun).

[15] The reverse rotation of Venus with pole-reversal beyond >90E, and Uranus rotation with a pole-reversal beyond >90E- is due to polar-drift similar to Earth's wobble-tilt of the axis of rotation; and

[16] Earth's polar-drift with wobble of the axis of rotation (that no one has a clue why…), found by IERS (International Earth Rotation and Reference Systems Service: see Table-6).

[16.1] J. X. Mitrovica et.al. "Mechanisms for oscillatory true polar wander" (see Nature 491, 244-248, Nov 08, 2012) and "How Earth's wandering poles return home" (see Physicsworld.com, Nov. 12, 2012).

The expert reports that the polar wander is the relative movement between Earth surface and its axis; and the computer model predicts the true polar wander that Earth surface tips over and then returns to its original position.

"If I sit at the pole, I see the pole shift up to 50° and then turn around, a process of oscillatory true polar wander.. I am really surprised that it took about 10x10^6 years to pull and push the poles."

When rock cools in a magnetic field it records the magnetic properties of the field; and millions of years later the magnetism in the rocks are decoded in the lab to measure changes in the orientation of the Earth magnetic field stored in ancient rocks. That shows the effects of the oscillatory polar wander.

Cause: See a message of Nov. 20, 2012 to Prof. Mitrovica-
I read the Abstract in Nature on your work- Polar Wander. I also read your comments in PhysicsWorld.

I agree it is an important topic to study. I am sure, you are aware of IERS work or measurements on Earth polar wander (including wobble of the axis).

I am an old retired engineer; and I spent lot of time on this topic. I have a complete solution to the topic, or should I say 100% solid proof that there is NO polar wander.

Yes, pole reversal is the Reality every 4.32x10^6 years- precisely. I will be happy to send you my calculations if you agree to give a careful review and willing to publish it in Nature.

I am sure we can make that deal in the interest of Man and Science.

With your permission, I can also say this- "If you sit in Boston or at the pole for 10 million years, Boston or the pole would not wander 50-70 degrees and then return close to that location".

And, your concept that the "bulge" and "Plate-stresses" counter the polar wander and make the pole return close to its original position is FLAWED.

That's not the Reality.

Please keep in touch- so, we can learn the Reality.

The simple equations and analysis discloses Neutron is the <u>cause</u> for the above [11-16.1].

The analysis shows Neutron force makes Sun to orbit-rotate, and Sun force, created by its rotation, makes the hot matter (the ejected plasma from sunspots) to orbit[9]; and the hot matter makes planets to orbit-rotate. The orbit-speed of the hot matter decreases with increase of the distance from the Sun due to collisions between the hot and cold matter.

As the distance between Sun-Planet, Rpa (or Planet-Moon, Rma) increases Sun force on the Planet, *Fspa (Planet force upon the Moon, *Fpma) decreases (see Table 2), and Neutron force on the Planet, *Fipa (upon the Moon, *Fima) increases in Z-direction. The analysis show that an increase in Neutron force on the Planet (Moon) causes an increase in orbit-inclination to the elliptic*[16-16.2] (like Pluto at ~17 degrees or Moon at ~5 degrees). Likewise, for Jupiter, Saturn or Neptune, as the distance between Planet-Moon increases Planet force on the Moon decreases, and Neutron force on the Moon increases. The analysis shows Neutron force on the Planet or Moon is the cause of orbit-inclination, and that is the cause of retrograde (polar) orbits to the outer Moons while the inner

Moons closer to the Planet keep orbiting in the equator plane in the direction the Planet rotates with little orbit inclinations.

*Force on the object acting on its surface creates a force-couple

We will <u>never</u> be able to find answer to the mystery or able to understand the structure of our-universe, if we keep going in the same path and refuse to take the indirect-clues left by Kepler and Newton.

NASA, in its website on gravity probe-B, states-

"If the GP-B results disagree with Einstein's theory "physicists may be faced with the challenge of constructing a whole new theory of the structure of the universe and the motion of matter".

Neutron, an invisible-insensible dark Object, is the cause for the mystery- not the Sun or Star. Newton gravity Eq.- modified to satisfy Kepler concept- is the clue for constructing a whole new theory of the structure of our-universe (the Atom) and the motion of matter.

Methods:

yetre dum vishraantam sa paramaanurithi bhaavaha

That part which remains after being cut repeatedly is known as 'atom'
(Kanaada's '*Vaisheshika Darshana*')

[Use of alphabets in the Eqs. will make it easy to understand the Text & Figures ...]

We utilized the observed NASA data on Planet orbits (Table 1), Earth-Jupiter-Saturn wind speeds (Table 5), Moon orbits, and corrugated rings etc... We made the Neutron with a mass and distance from the Sun active in a two-body problem of Kepler, and turned that into a three-body problem. We modified Newton equations and solved the three-body problem.

We show Neutron as the cause for the Planets to orbit around the Sun at varied distances from the Sun and at varied orbit inclinations to the elliptic plane in which Planets orbit. The analysis verifies the observed NASA data in all respects. That proves Neutron is the <u>cause</u> for the motion of matter in our-universe, an Atom, similar to Helium atom He. That must be true of Chadwick's neutron in a living atom, H.

We deduced Neutron mass, *Im ~4×10^{30} kg., and its' average-distance from the Sun, Rs ~161×10^{11} m. We utilized a trial-error method with various combinations of mass, Im, and distance, Rs, to come up with this right combination that satisfied the observed NASA data. The distance, Rs, varies with time-space, like the distance between Sun-Earth (1.496×10^{11}m=AU); and the ratio, Rs/AU ~$108*^3$.

*See Chadwick[10], here Sun has a companion, so Im ~1.006 x 2 (Sun mass).

We deduced the variable force-constants of Neutron (Bii, Bis), Sun (Bss, Bsi), and Planets (Bpp, Bpa). The force-constant can be construed similar to unit-pressure (w) or hydrostatic force (hsf) of water on a submerged-object, considering suns, planets, moons etc... are submerged, and floating in electro-magnetic field*4 (emf) of a fluid-ball. The emf and hsf forces are of the same species- where the emf causes gravity, and gravity causes hsf.

The proof that Neutron is the <u>cause</u> for the motion of matter in our-universe goes like this: Neutrons' cold (T~0K) and Suns' radiation (T~5777K at photosphere or T~1×10^{6}K at corona) disperses into space through the fluid like dark matter* or *params* within and beyond the fluid ball or Atom. Suns' hot matter dispersed within Neutrons' cold matter and appears in the form of electro-magnetic waves* of decreasing frequency or increasing length as the radiation, T, in space decreases with the increase of distance from the Sun. Neutrons' gravity attracts Suns' hot matter, and Suns' gravity attracts Neutrons' cold matter; and, the electro-magnetic force band (Figures 1-1c) between

Neutron-Sun keeps them together at a distance. Here, the rotation of Neutron makes the cold matter to orbit-rotate; and that causes the Sun to orbit-rotate. The rotation of Sun makes the hot matter to orbit-rotate; the orbiting hot and cold matter, together, make planets orbit-rotate. Planets attract hot and cold matter, and the rotation of the Planet makes the local hot and cold matter to orbit-rotate; the orbiting hot + cold matter cause the Moons and ring-objects to orbit-rotate.

*See Part 2 & 3 (will release after Part 1 takes roots).

We know (see [9, Figs.12-14]) that the 'sunspots erupt in low latitude bands on either side of the equator and drift toward the equator as each cycle progresses…; and the sunspot-groups extend in longitude, with one magnetic polarity associated with the leading spots (in the direction of rotation) and the opposite polarity associated with the following spots. … the erupting active region magnetic field diffuses at the equator…, and the higher latitude magnetic field elements transport toward the poles.' This activity, simply put- hot elements in S-hemisphere transport toward the magnetic-N under Neutron gravity, and the cold elements move toward the equator to diffuse; and the cold elements in N-hemisphere transport toward magnetic-S without much affect from Neutron gravity (Figure 1b- see bands & flow-lines). A similar flow or transport of the elements must be true in S-N hemispheres of the Earth (planets) under Neutron-Sun gravity; and must be true for the Moon under Neutron-Earth (Planet) gravity.

Hence, the rotation of Sun on its ZZ-axis disperses its radiation into space in XY-plane as the hot matter from Suns' interior continues to flow. The intensity of hot force and speed in the direction of its rotation will be high at the equator and low at the poles; the force lines spread out in spiral circles in XY-plane and as the distance from the Sun increases the intensity of the force and its speed decreases. We sense that hot force lines as a warm (hot) wind. The rotation of

Neutron on its YY-axis disperses its cold into space in XZ-plane, and the flow of cold matter from Neutron continues like at the Sun (assuming that to be the case). The intensity of cold force and speed in the direction of its rotation will be high at Neutron equator and low at the poles; the force lines spread out in spiral circles in XZ-plane and as the distance from Neutron increases the intensity of the force and its speed decreases. We sense that cold force lines as a cold (ice) wind.

The cold-force (or cold-wind) and the hot-force (or warm-wind) spreads out in spiral-circles at 90 degrees to each other within and beyond the sphere. When the hot matter speed increases abruptly it sucks the cold matter upward in spiral-circles and forms into a tornado on Earth or on the Sun, or as a cyclone in Ocean. The mean-intensity of hot or cold force at a point is inversely proportional to the distance from the emitter $\sim 1/R$ (R, distance of the emitter to the point). The resultant of cold-hot force at a point or at one of the emitters $\sim *B(1/R \times 1/R) = B/R^2$

*B=constant: represents electro-magnetic force in space

Newton Gp, gravitational constant of the Sun on Planet

We start with Newton Eq., to avoid re-inventing a wheel…

$$Gp = (Vp^2)(Rp)/Sm \qquad (1)$$

Gp=Newton gravitational constant,
Calculate using NASA's data in Table 1, for-
Vp= Planet orbit-velocity (m/s)
Rp= (Rpp+Rpa)/2 ~average-distance of Planet from Sun or orbit radius (m), where
Rpp=Planet perihelion distance from Sun (m), and
Rpa=Planet aphelion distance from Sun (m)
Sm = Sun mass (kg); and,

For <u>Moon</u>, replace Sm with Pm (Planet mass); Rpp with Rmp, & Rpa with Rma

Newton's Gp-value (see Table 2) is not constant at 6.673x10^-11 as believed*[5]. From Eq.1, we can see Gp varies from 6.6707x10^-11 at Mercury to 6.7866x10^-11 at Neptune with a rate of change, ~0.261x10^-24/m.

For Moon Gm= 6.734x10^-11

Evidently, Newton deduced the gravity constant, Gp, using Sun as the only source causing the Planet to orbit; therefore, Rp, in Eq.1 is the radius of a circular-orbit of the Planet around the Sun (neglecting Kepler ellipse). So, Newton Gp is an approximate value, and it is not useful to determining the nature of variable surface-gravity on a Planet, a reality that causes dynamic weather.

Planet force-constant, Bsp, due to Sun and Bip, due to Neutron

We assume Neutron rotates anti-clockwise about YY-axis in XZ-plane along X (see Figures 1-1c). It rotates in two ½ cycles of 10-11 years each having variable-areas of *polarity of magnetic-N in one hemisphere and magnetic-S in the other hemisphere with polarity switching sides in each ½ cycle or its magnetic N-S flip in 10-11 year cycles, and make Sun poles to flip[9].

*Polarity is associated with leading spots (+ or -)

That makes Sun orbit around Neutron about YY (XZ-plane) in X-direction or in the direction Neutron rotates. Per NASA Sun's axis of rotation-ZZ tilts ~7-7.5 degrees. That could be true only if Sun's axis of rotation-ZZ wobbles-spins, like Earth's ZZ. At this time, take Sun's ZZ tilts ~7.5 degrees with respect to *Indra*'s equator plane-XZ. So, due to this tilt of Sun's ZZ-axis, the force-band between *Indra*-Sun creates a force-couple on Sun's surface. That force-couple, acting on Sun's surface in XY-plane, rotates the Sun anti-clockwise about its ZZ-axis along X. In other words, *Indra* X-force (FisX) acting at the center of the Sun orbits the Sun while the difference in *Indra* X-force

(FisXt) at the top and (FisXb) at the bottom of Sun's surface, acting upon Sun's magnetic field flow, rotates the Sun anti-clockwise. That meant the net effect of the force-band between *Indra* and Sun would create a force-couple at the surface of the Sun due to the fact *Indra* X-force at the top of Sun's surface would be greater than the X-force at the bottom of Sun's surface. And, Sun's magnetic S-N flip once in 10-11 years with a change in the sunspots[9] polarity.

Planets orbit around the Sun about ZZ (XY-plane) in X- direction or in the same direction Sun rotates, and move in Sun's equator-plane. The distances between Sun-Planets and the orbit inclinations to the elliptic vary due to the variable nature of Sun XY-force and Neutron XZ-force. Planet rotates clockwise about its ZZ-axis in XY-plane or in opposite to Suns' rotation (Figure 1c). Earth's axis of rotation ZZ (N-S pole) continues to drift* toward the Sun, a little, in each orbit (or year) under Neutron X-force (the new-drift adds to the present tilt of the Earth at ~23.5 degrees). *See Table 6.

The inner Moons (and ring-objects etc..) orbit the Planet about ZZ (XY-plane) in X-direction or in the direction Planet rotates. Per NASA, Moon orbits Earth ~once in a month with little wobble and shows the same face towards the Earth. That is possible only if Moon is orbiting with its axis of rotation-ZZ (magnetic-S or N) pointing toward Earth's equator; so, Moon can rotate once a month under Neutron force. Moon orbit-distance depends upon Planet-Neutron XZ-forces; and its drift, if any, depends upon Neutron X-force. The orbit-inclination or retrograde (polar) orbits of outer Moons depend upon Neutron Z-force.

See **Figures: Atom Fig.9.1 to 9.1c, Sun orbits in XZ plane, Earth-Moon orbit in XY- plane.**

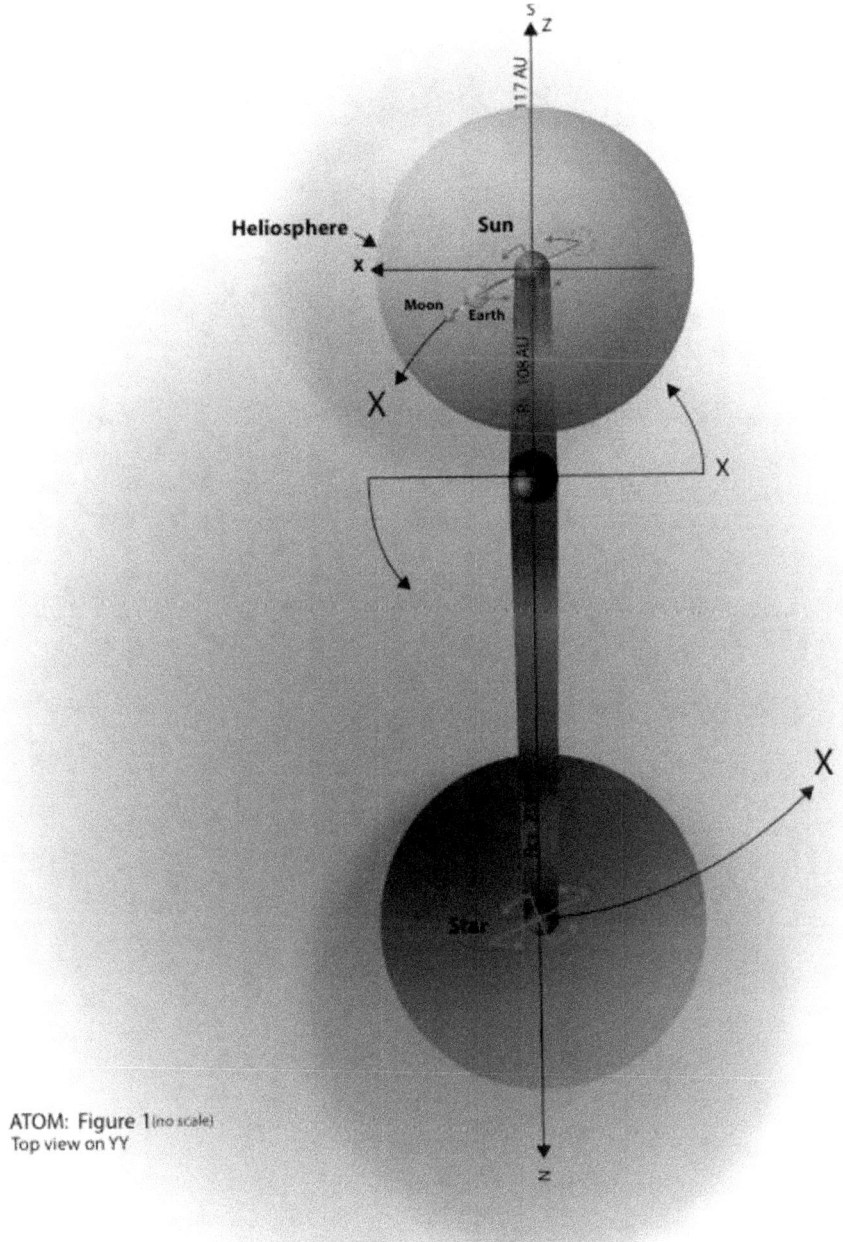

ATOM: **Figure 1**(no scale)
Top view on YY

205

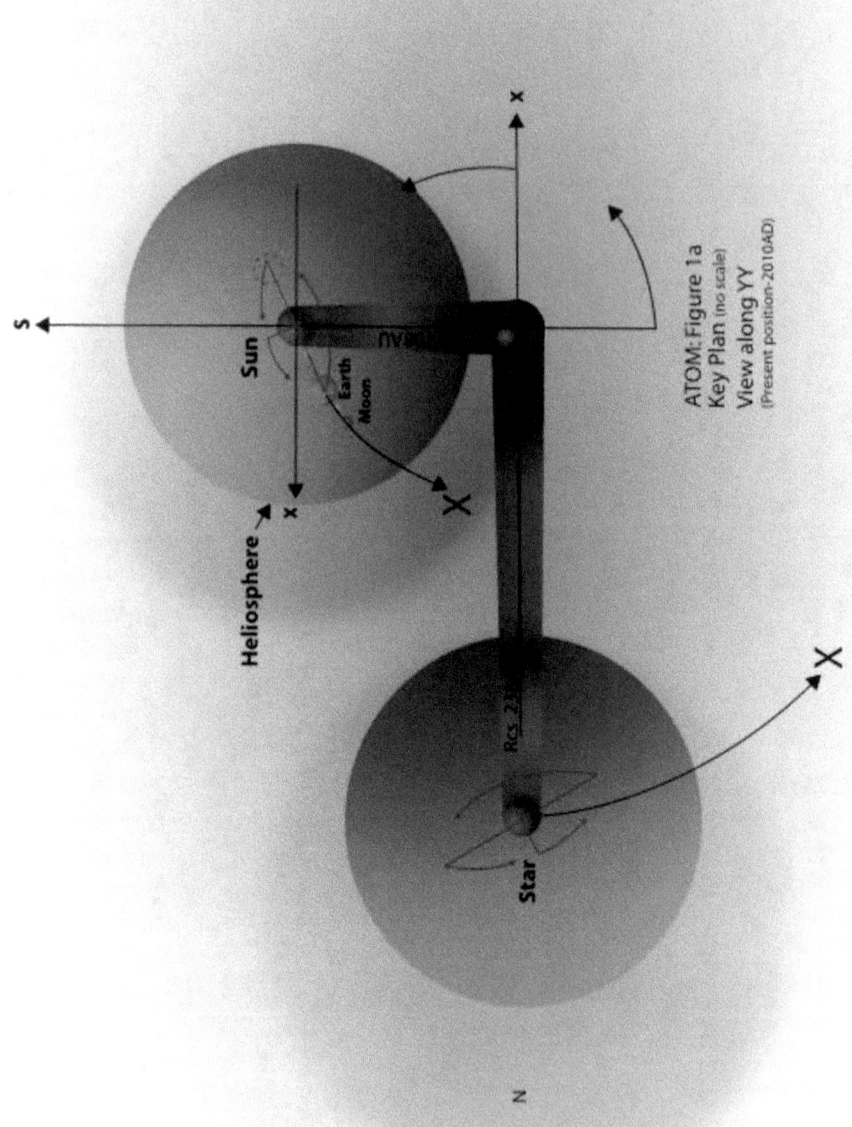

ATOM: Figure 1a
Key Plan (no scale)
View along YY
(Present position~2010AD)

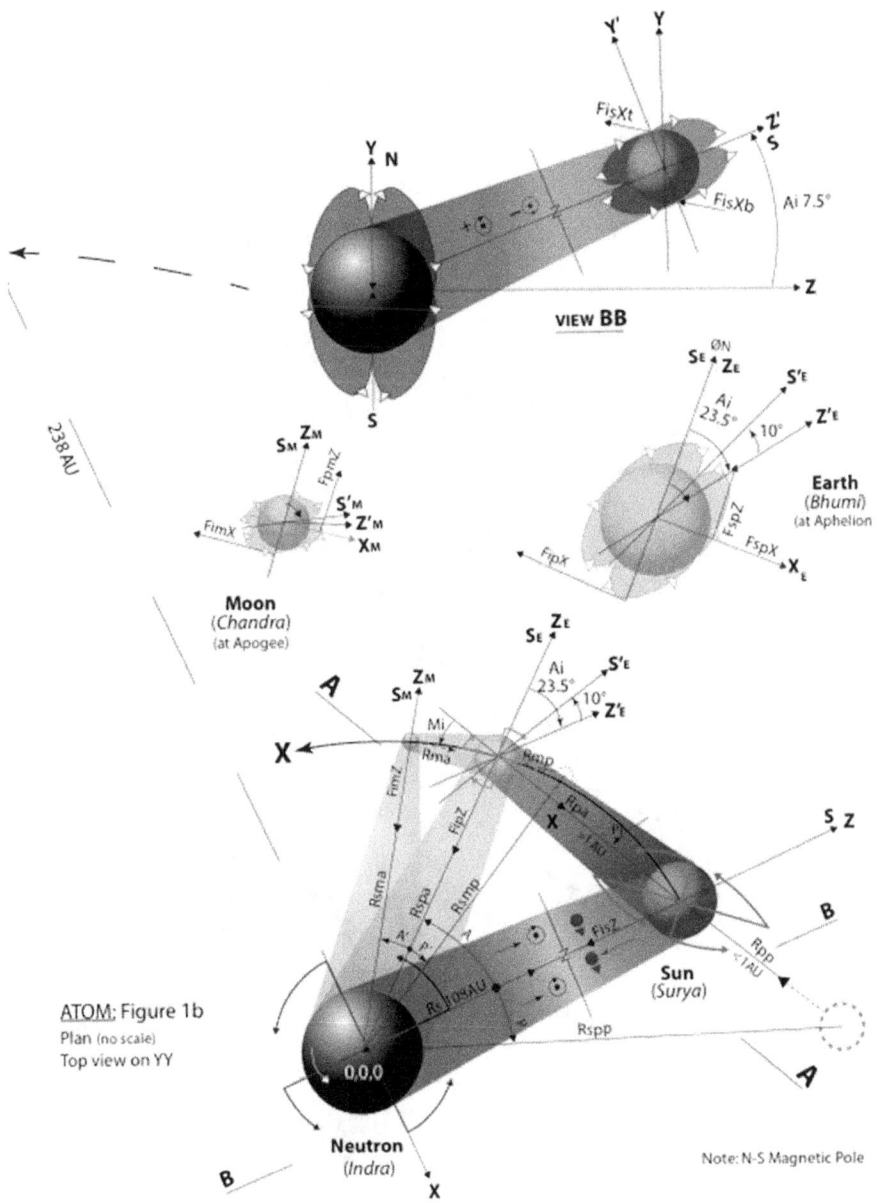

VIEW BB

Ai 7.5°

238AU

Moon
(*Chandra*)
(at Apogee)

Earth
(*Bhumi*)
(at Aphelion

Ai 23.5°

10°

Ai 23.5°

10°

Sun
(*Surya*)

Neutron
(*Indra*)

0,0,0

<u>ATOM:</u> Figure 1b
Plan (no scale)
Top view on YY

Note: N-S Magnetic Pole

207

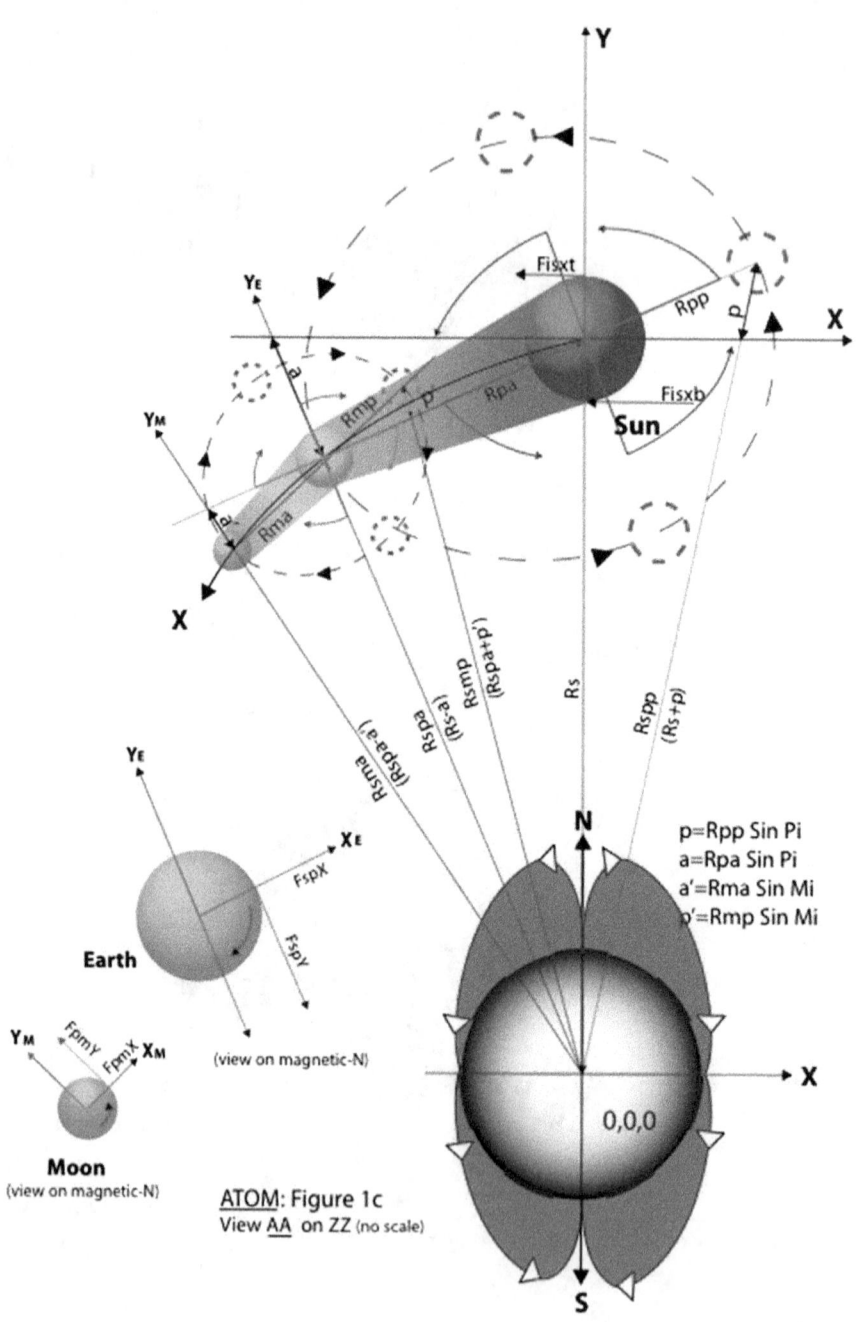

ATOM: Figure 1c
View AA on ZZ (no scale)

Under Neutron cold-force Sun(s) and Planets (Moons, ring-objects etc…) in our universe orbit, rotate, and spin at varied speeds with change in time-space. Analysis of the forces on a Planet at perihelion and aphelion (or on the Moon) and the motions can be determined using NASA data (Table 1) and the Eqs., below:

Suns' radial force on Planet, $\quad$ Fspp= Gp[(Sm)(Pm)/Rpp^2] $\qquad$ (2)

$\qquad\qquad\qquad\qquad\qquad$ Fspa= Gp[(Sm)(Pm)/Rpa^2]

Planets' radial velocity, $\qquad$ Vpp^2= (Fspp)(Rpp)/Pm $\qquad$ (2a)

$\qquad\qquad\qquad\qquad\qquad$ Vpa^2= (Fspa)(Rpa)/Pm

Or, from 2-2a we get, $\qquad$ Vpp^2=Gp(Sm)/Rpp $\qquad$ (2.1a)

$\qquad\qquad\qquad\qquad\qquad$ Vpa^2=Gp(Sm)/Rpa

From Eq.2, calculate Sun radial-force on the Planet at perihelion, Fspp (or Planet's radial-force on its' Moon at perigee, Fpmp etc..) knowing Gp from Eq.1 and Sm, Pm & Rpp from Table 1. Likewise, calculate Sun radial-force on the Planet at aphelion, Fspa (or Planet's radial-force on its' Moon at apogee, Fpma etc..). Then, from Eq.2a, calculate Planet orbit-velocity^2: Vpp^2, at perihelion, and Vpa^2, at aphelion (likewise Moon orbit-velocity^2: Vmp^2 & Vma^2). In Eq.2.1a, replace *Gp with +Bspp(or +Bsp), +Bipp(or +Bip) and use Rspp; and +Bspa(or +Bsp), -Bipa(or -Bip) and use Rspa to make both, Sun and Neutron active as below:

*For Moon +Bpmp(+Bpm), +Bimp(+Bim) & Rsmp;, and +Bpma(+Bpm), -Bima(-Bim) & Rsma

Vpp^2= [(Bspp)(Sm)/Rpp]+[(Bipp)(Im)(Sin P*)/(Rspp)] $\qquad$ (2b)

Vpa^2= [(Bspa)(Sm)/Rpa]−[(Bipa)(Im)(Sin A*)/(Rspa)]

*At equinoxes P=A=½(A+P); and for the <u>Moon</u> use Sin P', Sin A'

Newton Eqs., deal with a Planet orbiting around the Sun in one-plane; whereas, here, we need to deal with actions in two-planes at the same time: Planets orbit around the Sun in one-plane (Newton Eq.), and the Sun orbits around the Neutron in another plane (at ~90 degrees to Sun-Earth plane). In other words, we need to deal with a problem involving two orbits in two different planes that act together at ~90 degrees to each other. The simpler way to solve this problem is modify Newton Eqs. (of Sun-Planets in one-plane) to include the effect of Neutron on the Sun, Planets & Moons. Newton Eq. encompasses a principle that if an object is moving in a circular path at a given speed the centripetal force on the object relates to the mass and speed of the object. See Eqs.2 & 2a.

The term $(Bipp)(Im)(SinP)/(Rspp)$ in Eq.2b represents Neutron X-force on the Planet that acts toward the Sun (+) at <u>perihelion</u> as it increases force/speed of the Planet, or (–) at <u>aphelion</u> as it reduces the force/speed. The inclusion of orbit-inclination, Pi (or Mi) increases the length, Rspp (Rsmp), and decreases the length, Rspa (Rsma) which reflects Neutron Z-force on the Planet (or Moon); and at equinoxes, Rspp=Rspa=Rs. This means, the effect of Neutron X Z forces acting on the Planet at perihelion, aphelion are included in the modified Eq. 2b.

+Bspp(+Bsp) and +Bspa(+Bsp) represent Sun force constants at Planet perihelion and aphelion respectively; +Bipp(+Bip) and -Bipa(-Bip) represent Neutron force constants at Planet perihelion and aphelion respectively. Neutron force at perihelion is (+) since it acts to increase the force/velocity of the Planet (or Moon), and at aphelion it is (-) since it acts to decrease the force/velocity of the Planet (or Moon). See Figures 1b-c:

Add Eqs.2b, and get-

$$(Vpp^2)+(Vpa^2)=(Bspp)(Sm/Rpp)+(Bipp)(Im.Sin\ P)/(Rspp)+$$
$$(Bspa)(Sm/Rpa)-(Bipa)(Im.Sin\ A)/(Rspa)$$

Or simplify to get $=(\mathbf{Bsp})(Xsp)+(\mathbf{Bip})(Xip)+(\mathbf{Bsp})(Xsa)-(\mathbf{Bip})(Xia)$ (2c)

Then, solve Eq.2c using the values for Vpp^2, Vpa^2 (obtained from Eq.2a), and determine:

[+Bspp, Force constant of Sun at Planet perihelion,
+Bspa, Force constant of Sun at Planet aphelion] **Or**
Bsp, Force constant of Planet under Sun-force
[+Bipp, Force constant of Neutron at Planet perihelion,
-Bipa, Force constant of Neutron at Planet aphelion] **Or**

Bip, Force constant of Planet under Neutron-force

Here, Im =Neutron's Mass 4×10^{30} (kg)

   Rs =Average distance of Sun from Neutron 161×10^{11}(m)

   Sm=Sun mass 1.989×10^{30}kg

   Pm=Planet mass (kg)

   Rpp=Planet perihelion distance from Sun (m)

   Rpa=Planet aphelion distance from Sun (m)

   Rspp=Planet perihelion distance from Neutron (m)

      $= [(Rs)^2 + (Rpp)^2]^{\wedge} \frac{1}{2} + (Rpp)Sin\ Pi$

   Rspa =Planet aphelion distance from Neutron (m)

      $= [(Rs)^2 + (Rpa)^2]^{\wedge} \frac{1}{2} - (Rpa)Sin\ Pi$

   Pi=Planet orbit inclination to elliptic (degrees)

   Moon orbit-inclination, Mi (degrees)

Planet orbit inclination to elliptic, Pi (or Moon's, Mi) caused by Neutron Z-force makes Planet perihelion distance, Rspp (or Moon's, Rsmp, when Planet is at its aphelion) to <u>increase,</u> and aphelion distance, Rspa (or Rsma) to <u>decrease.</u>

For <u>Planet,</u> Rspp=Rs+(Rpp Sin Pi); Rspa=Rs-(Rpa Sin Pi)
When P=A=½(A+P), Rspp=Rspa=Rs
Where, Tan P=Rpp/Rs, and Tan A=Rpa/Rs

For example, looking at Figures 1b-c, it works like this- at Planet <u>aphelion</u>: as the distance between Sun-Planet, Rpa (or at Moon apogee the distance between Planet-Moon, Rma) <u>increases,</u> the angle A, between Neutron-Sun-Planet, and the angle Pi, the orbit-inclination to elliptic (or for the Moon angle A', between Neutron-Planet-Moon, and angle Mi, orbit inclination) <u>increases.</u> That change make the distance between Neutron-Planet, Rspa*, to be <u>less</u> than the distance between Neutron-Sun, Rs (or between Neutron-Moon, Rsma, less than the distance between Neutron-Planet, Rspa).

*At perihelion it makes Rspp>Rs.

An increase in Rpa (or Rma) decreases Suns' XY-force on the Planet (or Planet-force on the Moon); and an increase in angle Pi (or Mi) puts Planet closer to Suns' magnetic-N where the intensity of flow is greater (see Figure 1b, view BB), thus increases Suns' XY-force on the Planet (or Planet force on the Moon). Therefore, the result of an increase in Rpa and Pi could be a net-decrease in Sun force on the Planet, Fspa (or increase in Rma and Mi could result in a net-decrease in Planet-force on the Moon, Fpma); but the decrease of distance, Rspa (or Rsma) increases Neutron force on the Planet, Fipa (or on the Moon, Fima). The change in Rpa, Rpp (Rma, Rmp), Pi (Mi) & A, P (A', P') keeps Earth in orbit under Sun-Neutron forces (or Moon under Planet-Neutron).

We can determine Bsp, Bip from Eqs. 2b, 2c, and calculate the forces on the Planet (or Moon) using Eq. 2. We can calculate Sun XY force on the Planet, Fspa, Fspp (Planet's on the Moon, Fpma, Fpmp) and Neutron XZ force on the Planet, Fipa, Fipp (Neutron's on the Moon, Fima, Fimp):

For the Planet:

From Eq.2: Sun force, Fspa= Bsp[(Sm)(Pm)/Rpa^2]   (2d)

Fspp= Bsp[(Sm)(Pm)/Rpp^2]

Neutron force, FipaX= Bip [(Im)(Pm)/Rspa^2] Or

*Fipa=FipaX/Sin A

FippX= Bip [(Im)(Pm)/Rspp^2] Or

Fipp=FippX/Sin P

Neutron Z-force that makes orbit-inclination Pi (or Mi),

FipaZ=-Fipa (Cos A)*, and

FippZ=-Fipp (Cos P)

*At equinoxes A=P=½(A+P); so, Neutron Z-force= ½(FipaZ+FippZ).
The term Sin A in Eqs. 2b, 2c meant to include ONLY Neutron X-force (FipaX), the X-component of Fipa that acts in line with Sun X-force, Fspa; similarly, Sin P represents Neutron X-force (FippX) the X-component of Fipp. To obtain Neutron radial force Fipa or Fipp that acts on the Planet, the X-force be divided by Sin A or Sin P.

Sun-Neutron X-forces act in <u>line</u> ONLY at Planet aphelion, perihelion to keep Planet at a distance away from Sun:

So, at aphelion total X-force, $=\text{FspaX}-\text{FipaX}$ Contd:(2d)

At perihelion total X-force, $=\text{FsppX}+\text{FippX}$, and

Sun Y-force that orbits Planet aphelion (& rotates*downward view from Neutron),

$$\text{FspaY}=-\text{Fspa (Cos Ei)}$$

At perihelion (& rotates**upward view from Neutron),

$$\text{FsppY}=+\text{Fspp (Cos Ei)}$$

Or Fspa=FspaY=FspaX, If we neglect changing tangent-angle to the orbit (Ei)

Fspp=FsppY=FsppX, and

FspaZ=Fspa (Sin A)

FsppZ=Fspp (Sin Pi)

For the <u>Moon</u>: Eq. 2d works if we replace Sun with Planet and use Cos A', Cos P',
Sin A' & Sin P' plus terms: Rsmp=Rspa+Rmp(Sin Mi); Rsma=Rspa-(Rma Sin Mi); and Tan P'=Rmp/Rspp; Tan A'=Rma/Rspa

For the Planet as Rpa, Rpp increases A, P increases (for Moon Rma, Rmp, A' & P')- that meant Neutron Z-force increases on the Planet (or Moon) causing increased orbit-inclination, Pi (or Moon Mi) which could result in retrograde orbits to the outer Moons...as observed at Jupiter, Saturn.

Planet Mercury-Pluto: Bsp and Bip

Mercury-Pluto: @ <u>perihelion</u>- $(\text{Vpp}^2)+(\text{Vpa}^2)$ in Eq.2c being the total orbit-velocity^2 of the Planet caused by Neutron's radial-force (Sun + Neutron); it is to be split-up into Vsp^2, velocity^2 contributed by Neutron indirectly (thru Sun), and Vip^2, velocity^2 contributed by Neutron directly:

Eq.2c: $(Vpp^2)+(Vpa^2)=$**(Bsp)**$(Xsp)+$**(Bip)**$(Xip)+$**(Bsp)**$(Xsa)-$**(Bip)**(Xia)

$\quad\quad Vsp^2=[(Vpp^2)+(Vpa^2)](Xsp)]/[(Xsp)+(Xip)+(Xsa)-(Xia)]$ $\quad\quad$ (2P)

$\quad\quad Vip^2=[(Vpp^2)+(Vpa^2)](Xip)]/[(Xsp)+(Xip)+(Xsa)-(Xia)]$

And, from Eqs.2&2.1a,

$Fspp=[(Vsp^2)(Pm)/Rpp]=[(Bspp)(Sm . Pm)/Rpp^2]$

$Fipp=[(Vip^2)(Pm)/Rspp]=[(Bipp)(Im . Pm)/Rspp^2]$

$\quad$ We get, $Bsp=Bspp=(Vsp^2)(Rpp)/Sm$

$\quad\quad\quad Bip=Bipp=(Vip^2)(Rspp)/Im$

Mercury-Pluto: @ aphelion- $(Vpp^2)+(Vpa^2)$ in Eq.2c, is the total orbit-velocity^2 of the Planet caused by Neutron's radial-force; and, it is to be split up into Vsa^2, velocity^2 contributed by Neutron thru Sun, and Via^2, velocity^2 contributed by Neutron directly:

$\quad\quad Vsa^2=[(Vpp^2)+(Vpa^2)](Xsa)]/[(Xsp)+(Xip)+(Xsa)-(Xia)]$ $\quad\quad$ (2A)

$\quad\quad Via^2=[(Vpp^2)+(Vpa^2)](Xia)]/[(Xsp)+(Xip)+(Xsa)-(Xia)]$

And, from Eqs.2&2.1a,

$Fspa=[(Vsa^2)(Pm)/Rpa]=[(Bspa)(Sm . Pm)/Rpa^2]$

$Fipa=[(Via^2)(Pm)/Rspa]=[(Bipa)(Im . Pm)/Rspa^2]$

$\quad$ We get, $Bsp=Bspa=(Vsa^2)(Rpa)/Sm$

$\quad\quad\quad Bip=Bipa=(Via^2)(Rspa)/Im$

Neutron force *(Fipa, Fipp) and Sun force *(Fspa, Fspp) that acts on the Planet, in its orbit, varies from point to point, and cause Planet ZZ-axis wobble-spin. In Planets' orbit, at 2-locations (equinoxes) Pi=0, and A=P=½(A+P). It becomes Rpa=Rpp, and Rspa=Rspp ~Rs in Eqs.2P, 2A

*Neutron force on the Planet at aphelion, Fipa (or on the Moon, Fima), decreases at equinox, and continues to decrease to Fipp at perihelion (in ½ orbit); then, in the other ½ orbit, Fipp increases at equinox, and continues to increase to Fipa at aphelion. Whereas, Sun force on the Planet at aphelion, Fspa, increases at equinox, and continues to increase to Fspp at perihelion (in ½ orbit); then, in the other ½ orbit, Fspp decreases at equinox, and continues to decrease to Fspa at aphelion.

Note: In Eqs.2P & 2A, use the tabulated results in Table-2, and NASA data in Table-1, and calculate the forces Fspp, Fipp & Fspa, Fipa for each Planet; then, determine angle A, P, knowing Planet Rpa, Rpp (Rs=161x10^11m). Use the values in Eq.2d for each Planet and compare the orbit inclination, Pi, observed by NASA.

For the Moon, replace Sun with Planet, and calculate Fpmp, Fimp & Fpma, Fima, angle A', P', knowing Rma, Rmp, and Rspa etc. Use the values in Eqs.2d for each Moon, and compare the retrograde orbits observed by NASA.

From Eqs.2A & 2P we can determine:

Bsp=(Bspp+Bspa)/2, Force constant of Planet for Sun-force is equal

$\qquad$ ~Newton's Gp (see Eq.1) $\qquad$ (3)

Bip=(Bipp+Bipa)/2, Force constant of Planet for Neutron (or Bim for Moon..)

Bpp=Bsp+Bipp, and,

Bpa=Bsp-Bipa,

Bp=(Bpp+Bpa)/2, Force constant of Sun on the Planet (accurate value)

Table 2 lists values obtained for Bp that compares to Newton Gp, **Bp**, Bpp/Bpa, Bspp or Bspa, (Bsp), Bipp/-Bipa (Bip). For Planets Mercury-Pluto: the values vary linearly (approx.).

The curves in Figure 2 for the force constants, Bsp, Bss to Bsi, and Bip, Bis to Bii show the values increase with decrease of Sun radiation, T, in space between Sun-Neutron; and, the electro-magnetic wavelength, X, increases as T decreases. From this data, we can calculate the force constants of Sun Bss, Bsi and of Neutron Bis, Bii as below:

• Mercury-Neptune:

Bspp increases (linearly) from 6.67179x10^-11 to 6.80251x10^-11 or at a rate,

Bspr=[(6.80251-6.67179)x10^-11]/(44.1363x10^11)=0.29617x10^-24/m; and,

Bsar=[(6.80251-6.67179)x10^-11]/(44.6705x10^11)−0.29263x10^-24/m.

The mean, *Bsr=0.2944x10^-24/m; and, at this rate:

See **Fig.9. 2: Curves between Sun-Neutron show AU v. K, X, Bsp (Bss, Bsi), Bip (Bis, Bii).**

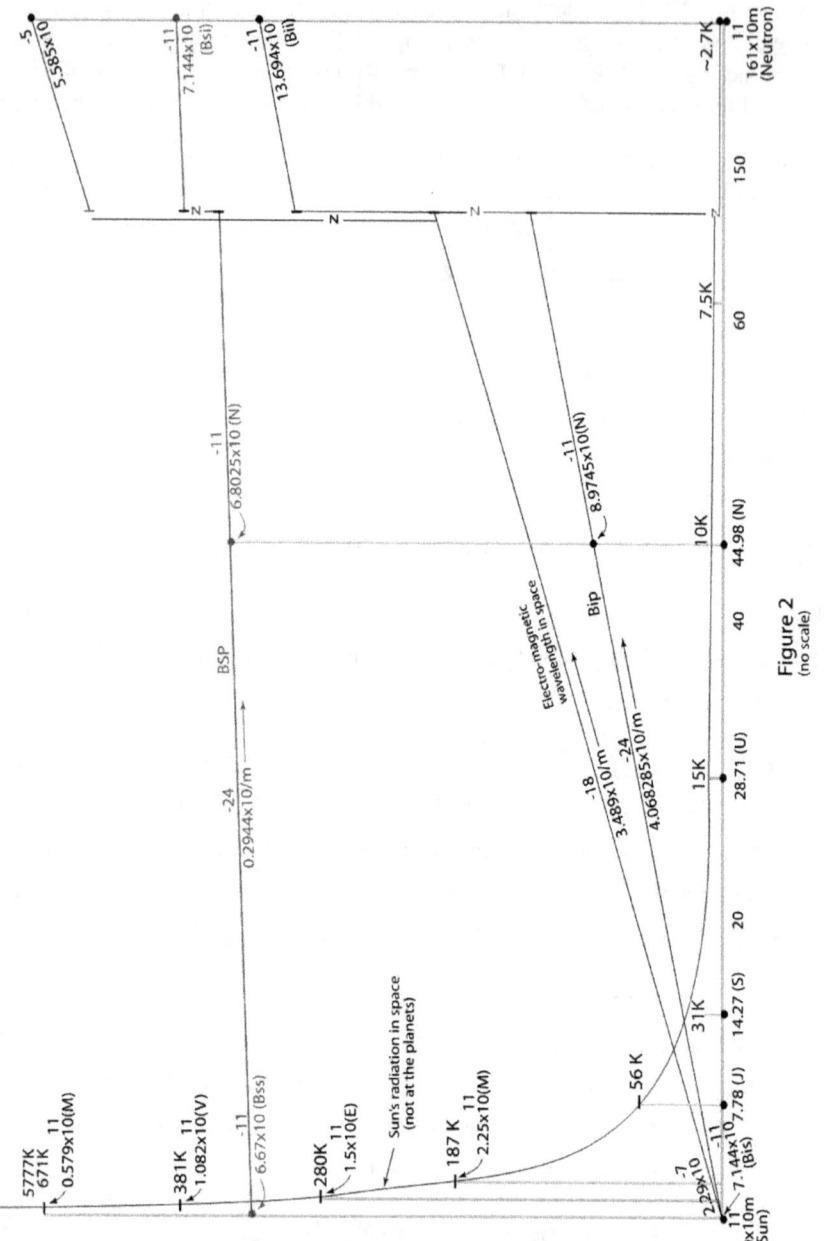

Figure 2
(no scale)

- **Bss**=(6.67179x10^-11)-(0.2944x10^-24)(0.5791)x10^11=**6.670x10^-11** at Sun (3s)
- **Bsi**=(6.67179x10^-11)+(0.2944x10^-24)(161-0.5791)x10^11=**7.144x10^-11** at Neutron
  Where, Bss & Bsi=variable-force constants of Sun.

- ## Mercury-Neptune:
Bipp increases from 1.9073x10^-13 to 181.5883x10^-13 or at a rate,
Bipr=[(181.5883-1.9073)10^-13]/(44.4034x10^11)
     =4.04656x10^-24/m; and,
-Bipa increases from 2.8933x10^-13 to 184.5037x10^-13 or at a rate,
Biar=[(184.5037-2.8933)x10^-13]/(44.4034x10^11)=4.09001x10^-24/m.
So, the difference in rate of increase, Bapr=(Biar-Bipr)=0.04345x10^-24/m.
The mean, *Bir=**4.068285x10^-24**/m; and, at this rate:
*Rate of increase same in all planes

- **Bis**−(7.144) = **7.144x10^-11** at Sun   (3i)
- **Bii**=[(7.144x10^-11)+(4.068285x10^-24)(161x10^11) =**13.694x10^-11** at Neutron
  Where, Bis & Bii=Variable force-constants of Neutron.

Results:
Taking the rate of change (Table 2 & Figure 2) is linear we can also calculate Bpp, Bpa & Bp from the equations below:

Bpp=[(Bss)+(Bsr+Bipr)Rpp] or [(6.670x10^-11)+(4.341x10^-24)Rpp]   (3a)
Bpa=[(Bss)+(Bsr-Biar)Rpa] or [(6.670x10^-11)-(3.7956x10^-24)Rpa]
  Bp=[(Bss)+(Bsr-Bapr)R] or [(6.670x10^-11)+(0.251x10^-24)R]   (3b)
Here, R=(Rpp+Rpa)/2, and Bp in Eq.3b is the mean-value ~ to Newton's Gp.

The rate of change or increase in Newton Gp at 0.261x10^-24 is equal to the rate of change or increase in Bp at 0.251x10^-24/m (see Table 2). The Bp or Gp will have no practical use hereafter. So replace Newton Gp with Bpp and Bpa to calculate the variable-gravity on the surface of the Earth (Planets) at any point in its orbit.

Gravity, $Gpp = Bpp(Pm/Pr^2)$ (4)

   And, $Gpa = Bpa(Pm/Pr^2)$

      $Bpp \sim Bpa$, at equinox

    Here, Pr = Radius of the Planet (m)

For gravity at any point in space replace Pr with $D = (Pr + R)$, where R = distance in (m) between Planet-surface and the point in space.

The variable-gravity at the surface of Earth is a primary cause for dynamic weather; and, the force constants Bii & Bis, Bsi & Bss, Bsp & Bip are dynamic in nature and change with time-space. The force-constant of Sun ($Bss = 6.67 \times 10^{-11}$) increases ($Bsi = 7.144 \times 10^{-11}$) between Sun-Neutron, with the decrease of temperature, T, in space or increase of distance from Sun; thus, the force of Sun on the Planet(s) increases at a flat-slope as the distance from Sun increases. Whereas, the force-constant of Neutron at Sun ($Bis = 7.144 \times 10^{-11}$) increases ($Bii = 13.694 \times 10^{-11}$) between Sun-Neutron at a steeper-slope.

Sun's force on Neutron, $Fsi = Bsi(Sm \cdot Im)/Rs^2$
Neutron's force on Sun, $Fis = Bis(Im \cdot Sm)/Rs^2$.
So, $Bis = Bsi$.

Looking in <u>Figure</u> 2, the force-constant of Neutron increases ($Bir = 4.068285 \times 10^{-24}/m$) at a steeper-rate than that of Sun (>14 times) between Sun-Neutron as temperature, K, in space decreases or distance from Sun increases. Thus, the force of Neutron on the Planet(s) increases at a steeper-rate as the distance from Sun increases or distance from Neutron decreases.

Neutron in cold, at all time-space, pulls (i) Suns-Planets towards it with magnetic-force; and in doing so pull (ii) Planets closer to Sun at perihelion and away at aphelion. Sun in hot, at all time-space, opposes the pull with electric-force (i) orbiting around Neutron; and in doing so pull (ii) Planets towards it. Planet replaces Sun in the case of Moon(s) and ring-objects.

Neutron provides the force (*Shakti*) to all and controls the suns, planets, moons and the rest in our-universe.

Our-universe (Atom) *v.* He (atom): Size & Work

Given *Indra* (Im) mass, ~4×10^{30}kg., and
2-SP mass, ~$2 \times 1.997 \times 10^{30}$kg..... SP=Sun+Planets
Mass-ratio, Im/2SPm=4/(2×1.997) ~1.0015
Given 2-Neutron (He) mass, ~3.349908×10^{-27}kg., and
2-PE mass, ~$3.3471078 \times 10^{-27}$kg...PE=proton+electrons
Mass-ratio, 2Nm/2PEm=1.00084 ~1.0015

Distance-ratio: our-Sun to (Earth)/(Neutron),
$$Dr= Rs/AU \sim \textbf{108*} \tag{5}$$
Force-constant ratio: (our-Sun)/(Neutron),
Fr=Bis/Bii ~**0.52***
*Universal-constants

Neutron force extends beyond our-Sun,
$$Ris \sim 108 + Bis/Bir \qquad \text{(see Eq.3i)} \tag{5a}$$
Ris=$108+(7.144 \times 10^{-11})/(4.068285 \times 10^{-24})$/m
   =$108+(175.60 \times 10^{11})/1.496 \times 10^{11}$ ~225 AU

We know (from Eqs.2P & 2A) Neutron is the provider of force or power to the Sun and Planets to move at variable-speeds. The speed of Sun (proton) or Planets (electrons) is proportional to its distance from Neutron (neutron). In a living-atom, the electron force level (or as physicist say ionization-energy) depends upon its distance from the nuclei' (proton-neutron); so is the Planet force level (to rotate, orbit, and wobble-tilt) which depends upon its distance from Sun-

Neutron. The proton force level (as physicists say nuclear energy) depends upon its distance from the neutron; so is the Sun force level (to orbit-rotate) which depends upon its distance from the Neutron.

Therefore, we can conclude that Earth force-level depends upon its distance from the Neutron similar to that of an electron force-level that depends upon its distance from the neutron.

For atom-He, we know the ionization-energy of 2-electrons as-
E1= 54.4eV and E2= 24.6eV
That gives a force-ratio: Er= E1/E2 ~2.21.

So, the distance-ratio,
Dr~(distance between electron2 to neutron)/(distance between electron1 to neutron) should be~equal to the force-ratio Dr~Er=**2.21***
*For atom-O, Dr~ 38.2 (Max.); for Cu, Dr~334.7 (Max.)

We know the mass-ratio: Im/2SPm ~ Nm/2PEm
So, like in He, the distance-ratio in our-Atom:
Dr~(distance between companion Sun to Neutron)/(distance between our Sun to Neutron) should be~equal to the force-ratio:

$$Dr=Rcs/Rs ~Er=2.21 \qquad (5b)$$

A careful review of Sky-photos from NASA, AAVSO.org, UCLA Galactic Center & Frog Rock Observatory show the space in our-galaxy is filled with-

- Binary-stars (**Astronomers agree binary stars exist in our neighborhood** [1&2.1]) with dark-spaces in between, like atom-He;
- Multiple binary-stars around a large dark-space, like atom-'C', 'O' or protein, with each star separated from the other, like in atom-He; and,
- Dark-space is much larger than the light-space where a star seen.

So we must **take** our Sun has a companion-Sun (**Astronomers fail to see that although they agree rest of the stars in our neighborhood are Binary-stars**), and the Neutron controls 2-Suns as shown in Figures 1-1a.
Use a distance-ratio, Dr~ 2.2

The distance between companion-Sun to Neutron,
Rcs ~108x2.2= 238AU
Neutron force extends beyond companion-Sun,

$$Ric \sim 238+238(Ris-Rs)/Rs \qquad (5c)$$
$$+238(225-108)/(108)$$
$$\sim 238+259 \sim 497AU$$

Our-universe is an elongated Sphere of diameter,
$$Di=225+497> \textbf{722AU}.$$

For the companion-Sun, the mean Bir2 ~Bii/Ric $\qquad (5d)$
Bir2 ~(13.694x10^-11)/(497x1.496x10^11)
~1.8418x10^-24 per m

Force-constant of companion-Sun for Neutron,
Bics=(13.694x10^-11)-[(238x1.496x10^11)(1.8418x10^-24)]
=7.14 x10^-11

We know the force-constant of our-Sun for Neutron when Suns are in opposite-side,
Bis =7.144 x10^-11

Force-constant of our-Sun for Neutron when Suns are on the same-side,
Bisc=(13.694x10^-11)-[(108x1.496x10^11)(1.8418x10^-24)]
= 10.718 x10^-11 or 1.5Bis

So, the mean force-constant of our-Sun, Bism ~8.931 x10^-11
Mean-speed of our-Sun, Vs^2= Bism(Im)/Rs or Vs= 4700 m/s[*6]
Sun's orbit period,   Ts= 6.2832(Rs)/Vs

$$Ts= (2160 \times 10^{\wedge}7)/(3.156 \times 10^{\wedge}7)= 684 \text{ years}$$

Neutron makes our-Sun orbit around it once every **684** years.

Speed of companion-Sun, $Vcs^{\wedge}2= Bics(Im)/Rcs$ or $Vcs= 2830$ m/s
Companion-Sun's period, $Tcs= 6.2832(Rcs)/Vcs$ or
$$Sco=(7899.44 \times 10^{\wedge}7)/(3.156 \times 10^{\wedge}7)$$
$$= 2500 \text{ years}$$

Neutron makes companion-Sun orbit around it once every **2500** years.

With the relative speeds and distances involved in each orbit around Neutron, the 2-Suns will be on the same side of Neutron once in every ~**942** years, separated by a distance of ~**130** AU. Or see in Figure 1-1a, our Sun-Planets and the companion-Star will be in opposite side of the Neutron every ~942 years.

The companion-Sun is visible from Earth when both Suns are on the same side of Neutron, and invisible when Neutron is between the Suns; rest of the time companion-Sun is partially visible (a *red-dwarf*) as Neutron and the dark-matter (physicist galaxy-dust[2.1]) being in our line of sight. Both Suns move in the same direction at different speeds; our Sun moves at a faster speed being closer to Neutron. And looking from the Earth that makes the companion appears as if moving in the opposite direction or backward.

So, the distance between our-Sun and companion-Star decreases in the first ½ cycle, and increases in the second ½ cycle, within the range of ~130 to 346AU. As our-Sun moves in its final ¼ lap (235 years) in a 942 year cycle, it catches up with the companion on the same-side of the Neutron- with the speed of Sun increasing, distance from Neutron decreasing, and the distance between Sun-Star decreasing from ~216 to 130AU. In the first ¼ lap in a new cycle, speed of the Sun decreases as the distance from Neutron increases, and the distance between Sun-Star increases from ~130 to 216AU...

Therefore, during the next ~200 years the sunspot activity*[7] may be effected as the Sun-Star line up on the same side of the Neutron, and Planets orbit closer to the Sun (as Bisc ~1.5Bis). That could make temperature on the Earth to rise, and the rising temperature melts the polar-ice at a faster phase as we witness at present. Hence, Sun must be in its final ¼ lap of its cycle, and moving closer to the companion with decreasing distance.

Astronomers conclude Proxima Centauri, a visible Star, is the closest to our Sun; and a recent-estimate of the distance between Sun-Star is set at ~4.22*[8] light years. The time-lapsed Photos of Proxima provided by S J Quirk of Frog Rock Observatory for the period from 1985-2005 show the Star has moved and still is moving (towards NE… with respect to Earth) as expected of a companion-Star to do. For that reason, Proxima is a good-candidate to be the companion-Sun.

Neutron binds our-Sun with the companion and with other Neutrons in our neighborhood of the galaxy; and our-Sun binds with Neutron and the companion-Sun... That makes our Galaxy a living-Cell, and it is bound*[9] together. The Neutron binds Suns, Planets, Moons and other objects in our-Atom into our-galaxy. The binding force is the **unified force of all fundamental forces** that appear in different forms [electromagnetic (casimir), gravity, weak and strong nuclear force, etc..], although of the same species.

It is the Neutrons (dark Objects)*[10], not clumps or halos, and not hundreds but more like millions (or billions) of various sizes exist in our galaxy, and make the dark matter (*params*) migrate back and forth between Neutron-Sun at various speeds. We sense the dark matter indirectly as Sun, Earth and other living inhale cold-warm matter and exhale warm-hot matter. Universe has <u>no</u> anti-matter or dark-energy; in fact, it is not a Universe. All life lives within

'Almighty-*Brahman*' and do 'duty'- like in a living being the creatures in the lungs, liver or kidney do the duty…

Per[9] Sun rotates once every 27-days,

*Vsr= [3.1416(2x0.6955x10^9)]/(27x24x60x60)

Suns' rotation-speed, = 1873.3 m/s say ~1,900 m/s

*Calculate Suns' mass that rotates:

From Eq.2, Neutron X-force on Sun,

FisX= Bis(Im . Sm)/Rs^2

=[(7.144x10^-11)(4x10^30)(1.989x10^30)]/(161x10^11)^2

=2.1927x10^24 kg

Sun orbits and rotates forward in the direction of force FisX

Force-couple that rotates Sun ~(FisX x Suns' radius).

So, (Vsr)^2 = (FisX x Suns' radius)/(Suns' rotating mass)

(1873.3)^2=[(2.1927x10^24)(0.6955x10^9)]/(Msr)

Suns' mass rotating, Msr ~ 4.34572x10^26 kg

To determine Suns' top layer thickness, t, that rotates,

t= (0.6955x10^9- r), and find r from-

[(1.4114x10^3*)(4pi/3)][(0.6955x10^9)^3 – (r^3)]=4.34572x10^26 kg

Solving for r, r=0.695396x10^9 m   *density kg/m^3
That gives, **Str** ~ (0.6955x10^9 - 0.695396x10^9) ~***0.11x10^6** m.

*This may be the effective top layer [corona to lower photosphere] because density of Corona, Chromosphere is low.

Table 1 provides Planet average orbit-velocity, Vp; distance between Sun-Planet, Rp (in XY-plane); and Sun mass, Sm. And, we know, Vp, is the total velocity contributed by Sun (Bsp) and Neutron (Bip).

Now, from Eq.1 and the data in Tables 1 & 2, we can split Vp back into its component parts, Vsp & Vip; and, determine Bhp, force-constant of Planet under Suns' hot-matter (see Table 3):

Where, $Gp= (Vp^2)(Rp)/Sm$, or

$Vp^2= Gp(Sm)/Rp$.   (see Eq.1)

So, $Vsp^2= (Vp^2)(Bsp)/(Bsp+Bip)$ (see Eq.2P, 2A)   (5e)

   $*Vip= (Vp)-(Vsp)$,

*Vip is mean-velocity at Rp=(Rpp+Rpa)/2, as Pi varies

Then, $Bhp= (Vsp^2)(Rp)/Sm$, (where, $Sm{\sim}1.989x10^{30}kg$)

Between Mercury-Earth, Bhp decreases,

$Bhme{\sim}[(6.648-6.611)x10^{-11}]/(0.917x10^{11})$

   $\sim0.04x10^{-22}/m$ (rate of decrease)

So, at Suns' surface,

$Bhs \sim[6.611+(Bhme)(1.496x10^{11}-*Suns' radius)] *0.6955x10^9 m$

   $\sim6.670x10^{-11}$ (equal to Bss, see Eq.3s)

Hot-matter orbit-speed at Sun surface,

$Vhsp^2\sim Bhs(Sm)/Sr$ (Sr=Suns' radius)   (5f)

$*Vhsp \sim 436,750$ m/s (233Vsr)

*Hot matter rotation-speed at Suns' surface is due to coronal mass injection from sunspot blast, plus Suns' rotation.
*Sun hot-matter and Neutron cold-matter collide at heliosphere 'nose' (HP), in the bands, and at corona to create temperature, T (See Figs. 1-1a).

The orbiting hot-matter at Suns' equatorial-plane and the orbiting cold-matter near the Sun collide*[2.1-2.3], and create the observed high-temperature, $T>1x10^6K$ at the corona. Such collisions make hot-matter lose its orbit-speed. And the hot-matter continues to lose speed as the distance from the Sun increases.

Hot-matter orbit-speed at Planet surface,
$$Vhp^2 = Bhp(Pm)/Pr \tag{5.1f}$$
Where, Pm=Planet mass, kg; and, Pr=Planet radius, m.

As the Planet rotates the orbiting hot-matter (coming from Suns' rotational-force) and the orbiting cold-matter (coming from Neutrons' rotational-force) collide and mix into hot-cold matter as the Planet rotates, and the mix loses orbit-speed as the distance from the Planet increases.

Per [9] the intense sunspot activity on either side of Suns' equator occurs between 22N-22S (equator remaining at 90 degrees to its axis of rotation ZZ), with a 2-week solar-Maxima followed by a 2-week solar-Minima in a 27-day rotation period. Suns' magnetic-S (that points away from Neutron) flips first to its N-hemisphere (and points toward Neutron) at ~ ½ year prior to the end of sunspot Maxima, in a 10-11 years cycle; and that event causes change in polarity of sunspots. Then, ~1½ years later, Suns' magnetic-N appears in its S-hemisphere. That shows Suns' S pole flips every ~21 years to point toward the Neutron..., nearly at the end of sunspot Maxima. So, Neutrons' specific cold-Magnetic force line must be the trigger[2,1] for Suns' pole flips while the cold force lines[2,3] keep the Sun in its proper orbit.
*IBEX experiment's 'ribbon-knots' created by hot-cold matter could be a sign of the trigger for pole flips?

Neutron to cause Suns' magnetic-S to flip every ~21 years is the clue that Neutron controls Sun and other objects in our-universe. We know Suns' orbit-period, So ~684 years; during this time Sun orbits 360 degrees around the Neutron at a radius, Rs ~108 AU.

Therefore, in 21 years Sun orbits a distance (in XZ-plane),
$$d = Rs. A,\qquad\qquad(5g)$$
Where, A= 360(21/684) ~11.0526 degrees
In 21 years, Neutron trigger-line must rotate a distance,
$$d = (2 \times 3.1416)\ Rz + (Rz. A)\qquad\qquad(5.1g)$$
Where, Rz= distance from center of Neutron to the trigger-line on ZZ-axis
Solve Eq.5g & 5.1g, we get, Rz=68.85AU*[2.1]
*IBEX experiment could not determine Rz.
Pluto Rpa at 49.3AU penetrates ~10AU into the trigger-line with greater Pi, e.

Neutron rotation (in XZ-plane) at Rz=68.85 AU (~39.15 AU from the Sun) can define the location for the critical-speed of 'cold-hot matter trigger-line':

$$*Viz = (6.2832\ Rz)/(4.4268 \times 10^{-3})\qquad\qquad(5h)$$
*Assume fluid-flow rotation-speed at the poles ~0

And, $4.4268 \times 10^{-3} = [(21 \times 365 \times 24 \times 3600)/1.496 \times 10^{11}]$
$$= (6.2832 \times 68.85)/(4.4268 \times 10^{-3})$$
$$\sim \textbf{97,700}\ \text{m/s or } 35.172 \times 10^{4}\ \text{km/hr.}$$

As the cold-Magnetic force lines disperse out along Z in XZ-plane, and if we set the cold-matter orbit-speed at 497AU (beyond c-Sun) equal ~0:
Cold-matter speed **decreases** at a rate,
Vciz~(97700)/(497-68.85)]= 228.2 m/s per AU

So, cold-matter orbit-speed at Neutron,
Vcip=97,700 +(228.2×68.85)=**113,450**m/s

At Sun, Vcis~ [Viz - (Vciz)(AU*-68.85)] m/s (5i)
*Distance from Neutron in AU

Therefore, cold-matter orbit-speed at Sun,
Vcis~ 97,700-(228.2)(39.15)   =**88,800** m/s

So, cold-matter orbit-speed at c-Sun,
Vcics~ 97,700-(228.2)(169.15)=**59,100** m/s

Neutron gravity,
*Gi=(Bii.Im)/(Id/2)^2
$\quad$ = (13.694x10^-11)(4x10^30)/(0.75x10^11)^2 [take Id~AU]
$\quad$ = 0.0974 m/s^2
*Likewise Sun gravity, Gs= (6.7x10^-11)(1.989x10^30) /(0.6955x10^9)^2 =275 m/s^2

Radiation in Space, K v. X Peak of the Wave

Experimental data on Earth show radiated heat energy varies with
wavelength and temperature; and, with increase in temperature
of the black body the peak of radiation wave shifts to shorter
wavelengths. That shift obeys a relationship under Wien's law.
Here, we modified that relationship to satisfy the characteristics
of dark matter in space or to specify the rate at which Suns'
radiation, K (Kelvin), disperses*[11] (absorbs-emits) between Sun-
Neutron (in XZ-plane), or Sun-Planets (in XY-plane) and beyond
as below:

Peak of radiation wave, Xmax.= (Kc)(2.898x10^-3)/K, m $\qquad$ (6)
$\qquad$ Or K.Xmax= 153.594x10^-6
Where,
Xmax= Wavelength at which radiation curve peaks, m
*Kc= Radiation disperse constant towards Neutron, **5.3x10^-2**
K= Absolute temperature of Sun's radiation in Kelvin
*If T~5777K per NASA, or if T~1x10^6K at Sun corona
(heliosphere nose (HP) & magnetic-N near Neutron),
.Xmax=**165x10^-6**, Kc=**5.6936x10^-2**

As temperature*[12], T in K, decreases with an increase of the
distance from Sun, the peak of the radiation wave shifts to longer
wavelengths at an average rate, Xs=**0.34891x10^-17**m/m (or

0.376105x10^-17m/m). See <u>Table</u> 4, Figure 2. The intensity of our-Suns' radiation could become Low by the time it reaches close to our-Neutron; and, as seen in NASA's WMAP map, there are an infinite locations where the observed T ~2.7K. Since the observed data was from a station in space located near the Earth, the record could be at distant-Neutrons of the light received from distant-stars, heliosphere nose (HP), corona, and the band at the Neutron. This shows that 'something' in space disperses radiation as it travels from the Sun or Star to the Earth and Neutron.

Apparently, the cold-Magnetic and hot-Electric forces act as 'tension and/or compression bands or springs' between Neutron-Sun-Planets (Earth) and transfer positive and negative forces and make the 'tie-in' between Planets-Sun-Neutron. The T~2.7K must be acting as a nerve to communicate*[13] between Sun-Neutron-companion Star and into the galaxy.

Therefore, this ~2.7K cannot be the cosmic microwave background (CMB) left over from a big Bang as hypothesized by physicists, although gamma ray-bursts (GRBs) or the Supernovae[2 & 2.1] do occur when the Atom (like our-universe) fuses with another Atom or splits. Neutrons exist, but not the black Hole as hypothesized. The cold-Magnetic (positive) and hot-Electric (negative) forces within a galaxy *[14] and between the galaxies must balance out, dynamically, as in our-Atom. Like the cytosol in a living-cell, the dark matter in our-galaxy must be 70% of the matter as the conductor of electro-magnetic waves.

We know ~4% of the matter is stars in the light-space, and knowing ~5% of the matter is Neutrons in dark-space, the rest ~21% of the matter must be the other insensible dark-Objects that act as Neutrons with DNA, RNA etc… to Create, Destroy and Recreate as required. We cannot sense the Creation that takes place within the Neutrons in space. That makes all creation in the Universe similar in basic structure inside to function, and varied from outside to make it difficult to say in what sense all living are alike.

The speed of radiation*[1.1] (light) on Earth at ~3×10^8 m/s cannot be a Universal constant- because, radiation passing through cold-matter in space slows wave progress as each particle interact with the radiation, T. The speed of radiation wave at a point of interest depends upon-
- Its initial speed at the emitter;
- Absolute temperature, T, at the emitter and at the point of interest; and,
- Force constant (or gravity) at a point of interest in space (Earth).

The change in temperature, T, and entropy gradient directly affects the red shift and the speed of light. The speed of radiated-light on Earth-

$$Ec=Sc(Ke/Ks)+(2Ge. \ AU)^{1/2} +Sw$$

The law $Sc/Ks=Ec/Ke$, comes from the general gas law:
$(P1.V1)/T1=(P2.V2)/T2$

Where, Sc=Radiation-speed at Sun at Ks=5777
Ec=3×10^8 m/s
Ke=280 (Table 4);
Ge=$9.8 m/s^2$;
AU for Earth=1.496×10^{11}m
Sw=29,648m/s (Table 3), Solar wind-speed near Earth
So, Ec= $Sc(280/5777)+(2 \times 9.8 \times 1.496 \times 10^{11})^{1/2} +29,648$

Speed of radiation at Sun,
Sc= $(2.98258 \times 10^8)(20.632) \sim \underline{61.5 \times 10^8}$ m/s

Law: Radiation-speed at any point,
$$Pc=Sc(Kp/Ks)+(2Gp. \ AU)^{1/2} \qquad (6.1c)$$
Where, Kp=Absolute temperature, K, at a point
Gp=Gravity, AU=Distance between Sun-point P, m

Frequency of light on Earth,
Ef=Ec/Ex =(3x10^8)/(0.55x10^-6) Ex=X (Table 4)
$\qquad$ = 5.455x10^14Hz, Visible range (400-700nm)

Frequency of light at Sun,
*Sf=Sc/Sx=(61.5x10^8)/(2.659x10^-8)
$\qquad$ =2.313x10^17Hz, X-ray range
*Frequency at Suns' corona (K~1x10^6)~ 10^22Hz, a gamma-ray

Radiation-speed at Neutron,
Ic=Sc(2.75/5777)+(2Gi. AU)^½
Light coming from Sun,
$\quad$ =(61.5x10^8)(0.000476)+(2x0.0974x160x10^11)½
$\quad$ = 4.70x10^6 m/s

Frequency of light at Neutron,
If=Ic/Ix=(4.70x10^6)/(55.85x10^-6)
$\quad$ =8.4x10^10Hz

Time between Sun-Neutron,
SIt=(161x10^11)/[(61.5+0.047)/2]10^8
$\quad$ =5.216x10^3 sec. or 1.45 hrs.

Note: The time between Neutron-Earth,
Tie=(161x10^11)/(Ic.+Ec)/2 ~107x10^3 sec or ~30 hrs.
Assuming Proxima (or any distant-star cleared 1AU dia. of Neutron, and it is
visible thru the dust[2.1], the radiation coming from the visible-Proxima would
display a speed, Ec~ c, near Earth.

The speed of light near the Earth, c~3x10^8m/s (per Eq.6.1),
whether coming from our-Sun or companion-Sun or a distant-emitter
or the reflected-light from Moon (or Mars etc…), must be related to
the temperature in space near the Earth ~280-295K.

Astronomers conclude Proxima, a visible Star, is the closest to our-
Sun; and a recent-estimate for the distance between Sun-Star is set at

~4.22 light years. Assuming our Sun and Proxima (the companion) are in the final ¼ lap of 942-year cycle, the distance between Sun-Star at present should be equal to ~216 AU. So, we can calculate a distance-factor to account for the speed-dilation due to Neutron and dark matter presence as below-

For Proxima, the correction factor:
Red-shift distance D/actual Da
=(*4.22x63239AU)/(216AU)=1235

For distant-star[2], the correction factor:
Red-shift distance D/actual Da
=(*5.8x10^27cm)/(1060000AU)**
=(3.877x10^14)/(1060000)=3.663x10^8

So, the rate of decrease in actual Da or a correction factor,
$$Df=[(1235)+(0.945x10^{-6})D] \qquad (6.1D)$$
Where, (3.663x10^8)/(3.877x10^14)~ 0.945x10^-6
So, actual **Da ~D/Df**,

*The distance based on Red-shift phenomenon within an Atom. To determine the red-shift distance, D (lyrs or cm), between Earth and a distant-star (in another-Atom) in space using change in red shift in that other-Atom is <u>incorrect</u>. We know the change of the red shift in our-Atom would yield a <u>correct</u> distance between Sun and an object within our Atom. Likewise, a change of the red shift in the other-Atom (if we know how to measure it sitting on our-Earth) may yield a correct result for the distance between that distant-star and its object (in that other Atom). Therefore, here we propose an approximate correction-factor to reduce the red-shift distance of a distant-star due to the existence of Neutrons/dark matter in the line of sight from Earth.

**Distant-star's X-ray flux ~5x10^-10erg/cm^2/s at 600km above Earth: see Fig.2[2] or Nature 476, p.42-4 (Aug.25, 2011). And Sun's X-ray flux ~5x10^-4erg/cm^2/s at 35,800km above Earth: see GOES15 data. Using X-ray flux of distant-star & Sun plus the data in <u>Table</u> 4, we can determine the distant-star's actual Da~106,0000AU & flare-temperature, T~60x10^6K.

Increase in wind-speed (Ws) v. (D) Distance decrease from Neutron

NASA did not know the reasons, but it recently (2009AD) found: The maximum wind-speed (Ws) on Saturn ~1,800 km/hr., and Jupiter ~600 km/hr.; NASA knew the maximum wind-speed (Ws) on Earth to be ~240 km/hr.

- On Earth, if the maximum wind-speed (Ws) is at ~240 km/hr. its surface-speed is ~1670 km/hr. at a rotation period of 0.99727 days.
- On Jupiter, if the maximum wind speed (Ws) is at ~600 km/hr. its surface-speed is ~45,260 km/hr. (~27xEarths') at a rotation period of 0.41354 Earth days.
- On Saturn, if the maximum wind-speed (Ws) is at ~1,800 km/hr. its surface-speed is ~35,535 km/hr. (~21xEarths') at a rotation period of 0.44401 Earth days.

If the mean force-constant of Neutron on Earth is normalized to 1.0 ~[0.5(Bipp+Bipa)=1.0, see Table 2], the mean force-constant on Jupiter will be 5.18 and on Saturn 9.52, as shown below:

Planet	Sun Distance Ratio	Neutron force-constant Ratio	NASA's wind-speed (Increase/AU)	Surface-speed	(Ss/Ws)
Earth	1.0 AU	1.00	240 km/hr. ..	1,670 kmhr	~7
Jupiter	5.2 AU	5.18	600 km/hr. (86)	45,260 kmhr	~75
Saturn	9.53AU	9.52	1,800 km/hr. (183)	35,535 kmhr	~20

Take NASA's wind-speed (Ws) is approximate, and it is the net speed at a location on the Planet for a mix of hot-cold dark matter. AU is the distance from the Sun to a given point.

We know (Eq. 5f) the orbit-speed of hot-matter at the Sun, Vhsp ~ 436,750 m/s in XY-plane (dispersing out along X, Y in spiral-circles). We know (Eq. 5i) the orbit-speed of cold-matter at Sun, Vizs~ 88,800 m/s in XZ-plane (dispersing out along X, Z in spiral-circles).

Neutron cold-dark matter (ISM cloud per IBEX) makes Sun to orbit in the direction the cold-matter flow; and Sun hot-dark matter makes the Planets to orbit in the direction the hot-matter flow. Therefore, wind-speed (Ws) on a Planet must be proportional to:
(1) Vip, orbit-speed of dark-matter (mix of hot-cold matter in XY-plane, Table 3) due to orbit-inclination created by Neutron force; and
(2) Ss, Planet surface-speed (see Table, above).

So, we can determine Ws at any Planet as follows:

$$\text{Assume, Ws} \sim \text{Wc(Vip x Ss)} \qquad (6.2)$$

From Table 3 & 5, for Earth,

$$\text{Ws=240km/hr. Wc} \sim \text{Ws/(Vip x Ss)}$$
$$= 28.92 \times 10^{-5}$$

Jupiter, Ws=600   Wc= 1.21×10^{-5}

Saturn, Ws=1,800   Wc= 3.5×10^{-5}

Knowing Wc, find Ws, for hot Planets,

$$\text{Ws} \sim 28.92 \times 10^{-5}(\text{Vip. Ss}) \qquad (6.3)$$

For cold Planets,

$$\text{Ws} \sim [1.21 + 0.529(\text{AU-5.2}) \times 10^{-5}](\text{Vip. Ss})$$

Therefore, we can conclude (see Table 5)-

- Neutron force on an object or Planet in space increases as the distance of the Planet from the Neutron decreases (see Table 2). The decrease of distance from Neutron increases wind-speed (Ws) on the Planet.

- The wind-speed (Ws) on the Planet is directly proportional to the orbit-velocity, Vip (under Neutron force), and Planet surface-speed, Ss.

The observed NASA data shows increasing wind-speed on Saturn over Jupiter and on Jupiter over Earth, but NASA is clueless why the wind-speed increases with an increase of distance from Sun. Thus, the observed data confirms existence of an Object or Neutron out there beyond Neptune that must be creating the cold and blowing cold-wind toward the Sun. In fact, the calculations show Neutrons' rotation is the cause of the cold-wind at speeds of $\sim35\times10^4$ km/hr. (at 39.15AU from the Sun).

Determine *Spin* or *Drift* of Earth axis of rotation-ZZ

Use IERS data (summary listed in Table 6)-

*Per NASA: "Earth's wobble.. is due to an ocean running hot and cold… due to fluctuating pressure on the bottom of the ocean, caused by temperature and salinity changes, and wind-driven changes in the circulation of the oceans…. But like a top, eventually the wobble should stop. This is what has puzzled scientists,…the wobble would have stopped…, unless some force was acting to keep it going… Aside from the long-term motions, the Earth's rotational axis and poles have two shorter periodic motions… (a free nutation…)…a yearly **circular motion**, and a steady drift toward the west caused by fluid motions in the Earth's mantle and on the surface. These motions are tracked by the International Earth Rotation and Reference Systems Service (IERS)…"

*Per IERS: ".. the angles which characterize the direction of the rotational pole within the Earth are called the polar coordinates, X and Y; and variation in these coordinates is called polar motion. The polar coordinates measure the position of the Earth's instantaneous pole of rotation in a reference frame which is defined by the adopted locations of terrestrial observatories. The coordinate 'X' is measured along the 0° (Greenwich) meridian while the coordinate 'Y' is measured along the 90° W meridian. These two coordinates determine the directions on a plane onto which the polar motion is projected…."

*Schreiber's "How to detect the… annual wobble of the Earth with.. Ring laser Gyro…" (PRL 107, 173904: 10/20/2011) and re-evaluate IERS EOP C04 combined series.

Method: Analysis of aphelion *Ra* and perihelion *Rp* per Table 6: see*[16]
1. In 1990: aphelion *Ra=0.5790* perihelion *Rp=0.0843*; *Ra-Rp~ 0.50*
   In 2009:   *Ra=0.5480*   *Rp=0.2204*; *Ra-Rp~ 0.25*

In ~20 years, *Ra* decreases ~5.5%, and *Rp* increases ~160% to become ~0.4*Ra*; and *Ra-Rp* decreases by 50%. The trend of <u>wobble diameter</u> (D=Ra+Rp) is upward and <u>increases</u> from 0.4252 to 0.7457arcsec/Yr.

2. Wobble* consists of 2 circular-shapes:
~mean *Ra*=16.0763/33~0.4848arcsec/Yr., Or 2piRa~<u>3.0461</u>; and
~mean *Rp*=4.91380/33~0.1489arcsec/Yr., Or 2piRp~<u>0.9355</u>

Therefore, the total circumference of 2-circular shapes=<u>3.98</u> ~4arcsec/Yr.
Assume 5% of that circumference is clockwise-spin of ZZ-axis at Ai~23.5E, Zs23.5=(4)(0.05)= <u>0.2arcsec</u>/Yr.*

*Wobble in 2 circular-shapes is being caused by *Indra* Z-force [acts at a mean angle 0-½ (A+P)-0] and Sun Z-force [acts at a mean angle Pi-0-Pi] in XZ-plane, see Figures 1b-1c. The estimate for Earth ZZ-axis spin is approximate [IERS data cannot measure the spin directly although IERS gives a polar drift ~0.00323arcsec/Yr. in IERS Fig.4]. Also, see IERS e-mail that shows Zs23.5 ~0.1506arcsec/Yr. (which equals to the mean- rate of spin/Yr for 180 degree spin).

——Original Message——
From: christian bizouard [mailto:christian.bizouard@obspm.fr]
Sent: Tuesday, April 28, 2009 8:10 AM
To: Mikki
Cc: christian.bizouard@obspm.fr
Subject: Polar Motion

Finally, I understood what you have computed: the pole coordinates differences $X_{i} - X_{i-1}$ at annual interval, for t_i = 1996, 1997, 1998, 1990, 2000
And what is your point? What's puzzling?

	X (arcsec)	Y (arcsec)	Polar Motion
Year 1996	-0.17	0.10	$[.17^2+.1^2]^{.5}$=0.197arcsec
Year 1977	0.14	0.10	=0.172

Year 1998	0.03	0.15	=0.154
Year 1999	0.10	0.10	=0.10
Year 2000	0.12	0.05	=0.13

(IERS Earth Orientation Parameter Center)

Observatoire de Paris.

61, avenue de l 'Observatoire 75014 Paris

From NASA data, we know, at present:

(**i**) Earth magnetic-S (N-pole) trails ~10 degrees (at 13.5E) behind its axis of rotation (Ai ~23.5E) and the Earth rotates at a speed of 465m/s.

(**ii**) Uranus magnetic-N (S-pole) leads ~59 degrees (at 157E) in front of its axis of rotation (Ai ~98E) and Uranus rotates at a speed of 2,590m/s (being Zero at Ai ~90E); and the speed is ~25% of Saturn rotation speed at 9,870m/s (Ai ~26.7E).

(**iii**) Venus axis of rotation (Ai ~177E) with magnetic-N (S-pole) close by, rotates at ~2m/s, or will become Zero at Ai ~180S.

(**iv**) Earth's magnetic field intensity is high at the poles and low at the equator. The intensity varies from ~25,000 nT at the equator to 65,000 nT (nano Tesla, 10^{-9}) or ~0.25–0.65 G (Gausses, 1G=100,000 nT). In the southern hemisphere (magnetic-N) the field points up and starts to incline downward as the latitude decreases until it become horizontal at the magnetic equator. In the northern hemisphere the field starts horizontal at the equator and inclines upward as the latitude increases until it turns down into the magnetic-S.

See Figure 1b: *Indra*, Sun, Earth and Moon, each one, has two-hemispheres (North, South). The magnetic field-flow indicates as if the two-hemispheres communicate with one another and do the work (like a living brain). See (**i**), (**ii**) & (**iv**) above (under Table 6). When the polar drift of the Earth gets to be Ai=90E, its rotation speed would become Zero, because the magnetic field-flow goes down into the Earth at the N-pole, and Sun magnetic field cannot create the Y-force (FspaY or FsppY) nor able to rotate Earth (the condition Uranus faced at Ai=90E). Again, see (i), (iii) & (iv) above. When the polar drift of the Earth gets to be Ai=180S, the rotation speed would become Zero, because the magnetic field-flow goes up (or out of Earth) at the S-pole, and Sun magnetic field cannot create the Y-force (FspaY or FsppY) nor able to rotate Earth (the condition Venus face at Ai=180S). That meant when Earth magnetic-S faces Sun at Ai=90E, the rotation and polar spin speeds become Zero; and when Earth magnetic-S flips to become magnetic-N and goes on to face *Indra* at Ai=180S, the rotation and polar spin speeds would become Zero. Notice when the magnetic-S is between Ai=0N-90E, Earth ZZ-axis leads until magnetic-S catches up to it at Ai=90E; when the magnetic-N is between Ai=90E-180S, Earth ZZ-axis should trail the magnetic-N (like at Uranus) until it catches up to it at Ai=180S. This is known as wandering magnetic poles…Although IERS data (Table 6) cannot directly provide a correct spin-speed of Earth ZZ-axis, we estimate at Ai=23.5E as Zs23.5~0.20arcsec/Yr.

Newton simply assumed that the force acts at the center of an object (or particle) for the purpose of his deducing gravity equations. In reality the X, Y, Z forces acting on an object transfer onto its surface as a force or force-couple and make the object (Sun, Planet or Moon etc…) to orbit, rotate, wobble-spin; and as the magnitude of the force or force-couple changes the speeds vary from point to point as the object orbits. For example, Sun XY-forces on a Planet depends upon the distance between Sun-Planet and the magnetic field intensity at the point where the Sun-Planet line crosses as the magnetic field-flow between magnetic-N to equator and equator to magnetic-S vary. Figure 1b shows hot and cold matter flows out at the magnetic-N, and cold matter flows toward the equator (as new-flows go in or out of the surface) until the flow turns back into the object at the equator; hot and cold matter flows out at the equator, and cold matter flows toward the magnetic-S (as new-flows go in or out of the surface) until the flow turns back into the object at magnetic-S. Thus, the magnetic field-flow will be <u>high</u> on both sides of magnetic-N and magnetic-S, and <u>low</u> at the equator.

At magnetic-N (Ai=180S, with ZZ-axis on N-S poles): looking towards 270W and towards 90E, we see a flow-intensity of <u>high</u> and <u>up</u> decreasing (0.65-0.25G) as it goes <u>downward</u> in a curved-path (or flow-inclination downward) to the equator with a <u>low</u> intensity and <u>horizontal</u>. At magnetic-S (Ai=0N, with ZZ-axis on S-N poles): looking towards 270W and towards 90E, we see a flow-intensity of <u>low</u> and <u>horizontal</u> at the equator increasing (0.25-0.65G) as it comes <u>upward</u> in a curved-path (or flow-inclination upward) to the magnetic-S with <u>high</u> intensity and <u>down</u>. Now, see how Earth magnetic field flow and its inclination work with *Indra* XZ-forces (FipaX, FipaZ or FippX, FippZ) and with Sun X-force (FspaZ or FsppZ). Figure 1b shows *Indra* X-force acts along X-axis (in XY-plane) on Earth-diameter YY; Sun X-force or its Z-component acts along Z-axis (at 90 to XY-plane) on Earth-diameter YY. Notice the difference: magnitude of *Indra* X-force remains the same (FipaX~ FippX), but Sun Z-force will vary from maximum at perihelion to "0" at equinox, and maximum at aphelion to "0" at equinox. This is due to the orbit-inclination, Pi (or the inclination of XY-plane: see Fig.1d, a view of *Swastika*...Earth orbit), at perihelion and aphelion, which become "0" at equinoxes. In other words, as Earth orbits Sun Z-force moves around the Earth once a year with its maximum shifting from perihelion to aphelion.

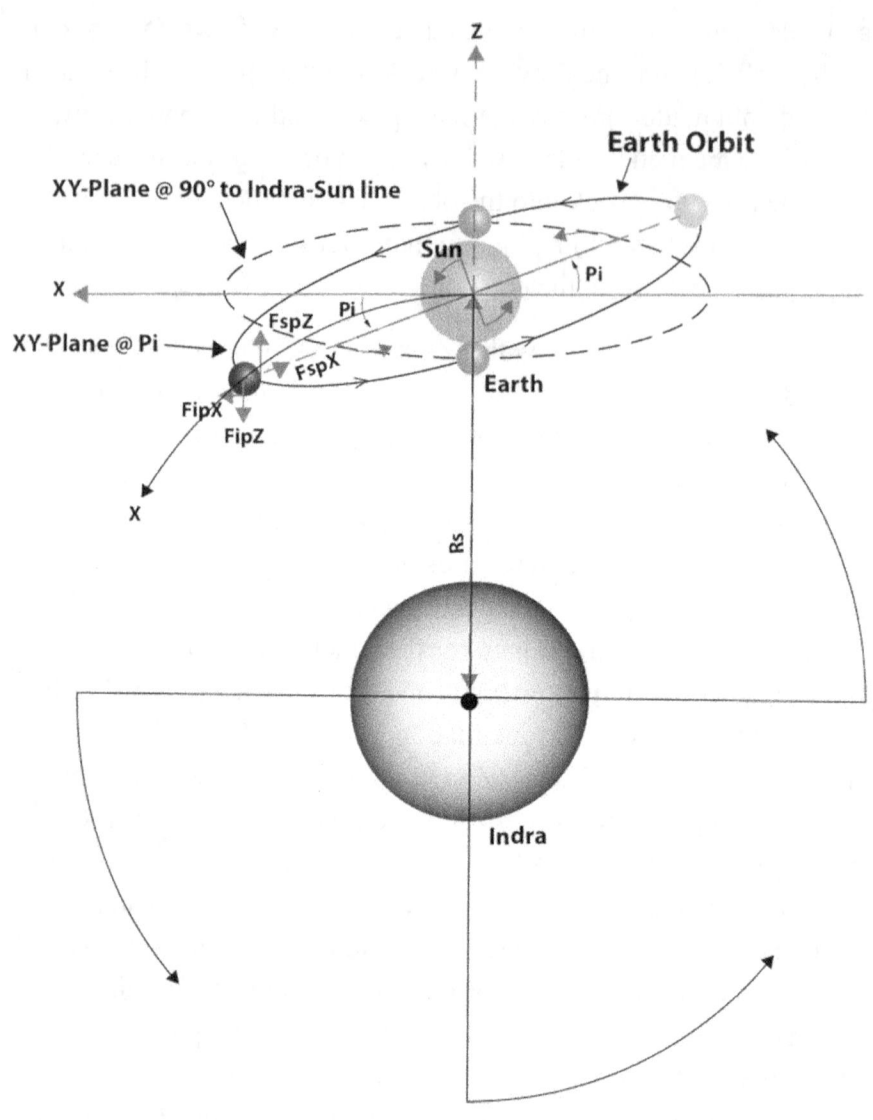

Fig.9.1d: A view of *swastikas* at Sun (XY) and *Indra* (XZ)

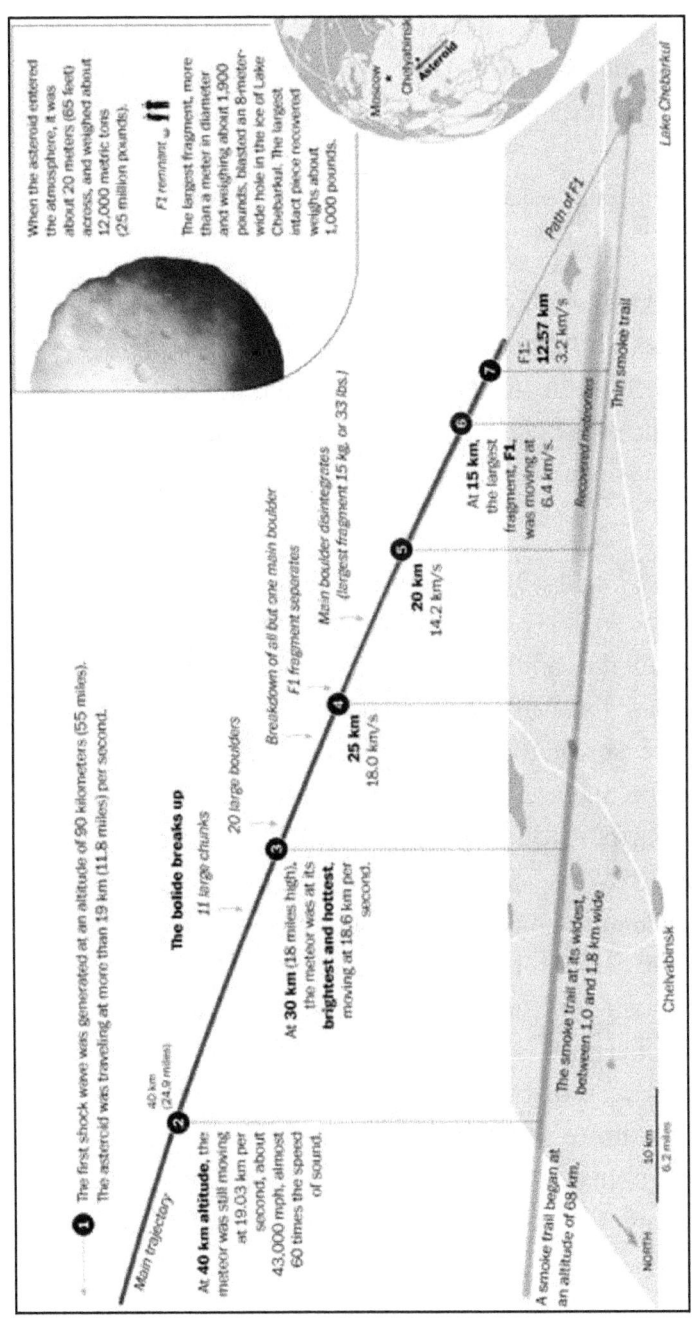

SOURCE: Nature
Fig.9.1e: Path of the Chelyabinsk asteroid* (on February 15, 2013 morning)
[Since the disintegration of an asteroid over Russia in February, scientists have compiled data from numerous sources — including YouTube videos — to gain a better understanding of the meteor's 500-kiloton release of energy].
Washington Post. *Patterson Clark*, Nov. 6, 2013

*American Musium of Natural History defines asteroid as a rocky body found mostly in the asteroid belt, between Mars and Jupiter. Jupiter is the largest planet in our solar system, and its gravity is very strong. Asteroids, which are much smaller than planets, are sometimes pulled out of the asteroid belt by the force of Jupiter's gravity. Many of these asteroids then travel toward the inner solar system—where they can collide with Earth.

If Jupiter can pull an asteroid out of Sun's gravity, that asteroid may only travel toward the Jupiter, not into the inner solar system—or if our Sun and Jupiter are in such a war, Sun would not leave Jupiter or other planets alone. Therefore, a more fundamental principle must be behind for the asteroids coming our way. See Fig.9.1d and Fig.9.1e:

In Figure 1d (note Earth's N-pole points to *Indra* in the present *yuga*), Earth orbits in XY-plane in the direction Sun rotates, and that plane is inclined at angle, Pi (Pe). This inclination of the orbit-plane is due to *Indra*'s swastika-force on the Earth. The intensity of that force increases on the earth (or planets or asteroids) between perihelion and aphelion as the distance decreases between *Indra*-Earth. And *Indra*'s force makes Sun to rotate anti-clockwise—the arrow on Sun's *swastika* shows the direction in which Sun rotates and planets (asteroids existing between Mars-Jupiter) orbit. The Sun's force makes planet or asteroid rotate clockwise or opposite to Sun's rotation. They orbit around the sun in XY-plane, inclined at different angles, Pi...Pn—depending upon its own mass and distance from the sun. In this process, it is most likely one or more asteroids bump into each other or collide (like Earth-Chelyabinsk asteroid, or see [8] where Saturn/Jupiter rings appear corrugated, looking along the equatorial plane, which are caused by the *swastika*-force of *Indra*... as the ring-objects move up or down like the Planets/asteroids, with varied orbit inclinations), and kicked in/out or up/down from its orbit in XY-plane. When that happens the asteroid assumes a new orbit distance from the sun and embarks into a unique orbit or trajectory of its own at a new angle, Pi, (to XY-plane) under Indra's force. Likewise, an asteroid orbiting the companion-Star could undergo a similar scenario, stray down heading towards our-Sun, and collide with Earth or a Planet—this scenario is more likely to occur during the next 200-400 years at which time the sun-star will be on the same side of *Indra*.

On February 15, 2013, Earth must have been at perihelion with its N-pole pointing down in the direction of *Indra* and Sun. See Fig.9.1d. On that specific day-time the orbit-paths of Earth-asteroid crossed as they were going in the same direction—Earth at ~30km/s and asteroid at >*30km/s (*varies with mass/distance). Asteroid speed was reduced to about 19km/s at ~40km above the Earth—due to the fact that it was travelling in opposite to Earth's rotation or against the wind.

The Figure 1e, shows an asteroid (weighing much more than 3,000 pounds) orbiting our-Sun at a speed of more than 30km/s, came from the southeast and heading to the northwest at an angle, Pi>Pe (or ~20 degrees to the XY-plane), crossed path with Earth's orbit and collided. It travelled a distance of ~110km to the northwest with its speed decreasing, and crashed into an icy lake. Using this data and the methodology described elsewhere in this Chapter 9, we may be able to estimate the trajectory, speed, and mass of the asteroid—thus, able to determine the mass of an asteroid most likely to hit Earth.

Clockwise spin of Earth ZZ-axis:

The spin at the present time, call phase 2, 0N-90E (or 180S-270W):
At magnetic-N (S-pole at Ai=180S) the magnetic field-flow is up
(YY) and *Indra* X-force acts along XX-axis (right to left). And,
when *Indra* X-force is able to push Earth forward slowly (to the
left) that motion will make the magnetic-N move with Earth, and
facilitate Sun magnetic field flow interact with Earth magnetic field
flow; thus, creates a new-ZZ axis about which Earth starts to rotate,
slowly. As a result, Earth ZZ-axis at N-pole (Ai=0N) moves forward
(toward 90E) and the magnetic-S follows.

At this stage, the magnetic field-flow at the equator (located at 90E
& 270W) will be low and horizontal; and the flow-intensity increases
along an upward curved-path to magnetic-S (Ai=0N) and turns down
into magnetic-S. Given that (see Fig.1d):

At aphelion, Sun Z-force and the upward inclination of Earth
magnetic field-flow (from equator to N-pole) will be in the same
direction; therefore, the interaction could create a low 'drag force'
opposing clockwise spin of Earth. And at perihelion, Sun Z-force
acts in opposite to the upward inclination of Earth magnetic field-
flow; therefore, the interaction could create a high 'drag force'
opposing clockwise spin of Earth.

Since Earth magnetic field-flow will be high and up at magnetic-N
(Ai=180S) and decreases along a downward curved-path to low and
horizontal at the equator (Ai=90E, 270W), the downward inclination
of that field-flow from 180S to the equator at 90E will be in opposite
to the direction of *Indra* X-force; therefore, the interaction could
create a high 'drag-force' to cause a faster spin of Earth ZZ-axis
clockwise (followed by magnetic-N).

And, by the time Earth ZZ-axis and the magnetic-N arrive at Ai=270W (from 180S), the ZZ-axis and magnetic-S arrive at Ai=90E (from 0N); then, the equator will be at 180S-0N with low and horizontal field-flow. In summary, *Indra* X-force (with high drag force) spins the Earth clockwise at all time, and Sun Z-force (low or high drag force) spins Earth anti-clockwise.

The spin in the past (or future), call phase 1, 270W-0N (or 90E-180S):
At equator (Ai=180S) the magnetic field-flow will be low and horizontal, and looking toward 90E it increases along an upward curved-path to high and up and turns down into magnetic-S (at 90E). *Indra* X-force acting along XX-axis (at Ai=180S) interacts with Earth field-flow (going toward 90E) in opposite to the direction of X-force; therefore, that creates a high 'drag force' and facilitate *Indra* X-force to pull the magnetic-S forward (which is stuck at 90E facing the Sun). At this stage analyze (i) & (ii) above (see under Table 6), and understand how or why the magnetic poles wander:
(i) shows in phase 2, the magnetic-S (N-pole) trails behind ZZ-axis as it moves forward, and keeps ZZ at or near high flow until the magnetic-S and ZZ unite at Ai=90E; that meant the magnetic-N (between 180S-270W), too, trails behind ZZ-axis (S-pole) and moves forward as *Indra* X-force pushes it from behind, and keeps ZZ at or near high flow until the magnetic-N and ZZ unite (at Ai=270W).
(ii) Shows in phase 1, the magnetic-S (N-pole) turns into magnetic-N (S-pole) once it can get out of Sun's grip at 90E, and starts to lead Earth ZZ (from Ai=90E to 180S); likewise, at Ai=270W, the magnetic-N (S-pole) turns into magnetic-S (N-pole) and starts to lead ZZ (Ai=270W to 0N).

That turning of magnetic-S into a magnetic-N (at 90E) is a pole-reversal. It results in a reversal of the magnetic field-flow between 90E-180S (and 90E-0N). In other words, the magnetic field-flow will be <u>high</u> and <u>up</u> at the new magnetic-N (near 90E), and the flow-intensity decreases along a <u>downward</u> curved-path to <u>low</u> and <u>horizontal</u> flow at the equator (180S and 0N). That meant the downward inclination of the magnetic field-flow (from 90E to 180S) and *Indra* X-force will be in the same direction; therefore, the interaction could create a <u>low</u> 'drag-force' resulting in a <u>slower</u> spin of Earth ZZ between Ai=90E-180S. In this phase, *Indra* X-force (with <u>low</u> drag force) spins Earth clockwise, and Sun Z-force (at <u>low</u> or <u>high</u> drag force) spins Earth anti-clockwise. In summary, when the magnetic-N trails behind Earth ZZ (between Ai=180S-270W or 0N-90E) *Indra* X-force <u>pushes</u> the magnetic-N forward at a <u>faster</u> speed, and when the magnetic-N leads in front of Earth ZZ (between Ai=90E-180S or 270W-0N) *Indra* X-force <u>pulls</u> the magnetic-N forward at a <u>slower</u> speed.

The rotation speed of Earth (at Ai~23.5E) under Sun Y-force will keep on decreasing in small-steps until its ZZ-axis drifts closer to Ai=90E. At Ai=90E, Earth would not rotate or rotates once in each orbit, under *Indra* force, and continues to do so for thousands of years (like our-Moon is doing, now, once in each orbit around the Earth). During that period, Earth orbits around the Sun showing the same face (with magnetic-S at aphelion and magnetic-N at perihelion) towards the Sun until magnetic-S (at Ai=90E) turns into magnetic-N, like at Uranus- a pole reversal. During this time, life on Earth could become sparse or none. And, there is <u>no</u> real fix to the decelerating rotation speed despite the 'leap second' fix. We, also, know Moon is moving away slowly as Earth rotation-speed is decelerating.

The wobble-spin of Earth ZZ-axis is <u>not</u> 'due to fluid motions in the mantle or on the surface'. The wobble-polar drift of Earth (mantle + crust=342km thick) is due to *Indra* Z-force that acts at 90 degrees to XY-plane at equinoxes (in –Z direction) since the <u>vertical</u> angle at aphelion-equinox-perihelion changes from 0 to ½(A+P) to 0, and due to Sun Z-force that acts at 90 degrees to XY-plane (in Z direction) as the orbit-inclination changes from Pi to 0 to Pi. The moving of Earth mantle + crust makes mountains, valleys with volcano eruptions, and earthquakes. The pole-reversal or magnetic S-N pole flips from one hemisphere to the other originate in Earth's core as the axis of rotation continues to spin either leading or following the magnetic-S or N. In the present ½ cycle (Ai=0N-90E), Earth began with a polar spin speed of 0 arcsec/Yr at Ai=0N, and the spin speed continues to increase until ZZ-axis arrives closer to Ai=90E; then it becomes 0 arcsec/Yr at Ai=90E. The spin speed cannot be determined in an experiment, like measuring wobble by IERS. The best we can do is to make an estimate for the polar motion using IERS data shown in <u>Table</u> 6, and compare the result using Eq. 2d, 2D. See Endnotes at *[16.1].

Based on the above, we are to determine the interplay of *Indra*-Sun forces upon Earth magnetic field-flow and calculate the speed of orbit, rotation and spin. The principle behind the cause for the wobble-spin is *Indra* Z, X-force and Sun Z-force acting upon Earth magnetic field-flow. *Indra*-Sun, together, cause the wobble of ZZ-axis that IERS data represents.

A summary of forces and motions:
(i) Sun X, Y-forces (FspaX, FspaY or FsppX, FsppY) act in XY-plane at 90 degrees to one another, and make the Planet orbit and rotate (on its ZZ-axis) in the direction Sun rotates. Sun X-force band

keeps the Planet at a distance from the Sun, and Sun Y-force band rotates the Planet.

(ii) Sun-*Indra* net X-force (FspaX-FipaX or FsppX+FippX) acts (approximately) in line only at Planet aphelion and perihelion, and keep the Planet at a distance from the Sun; rest of the time in its orbit, *Indra*-Sun forces do not line up, but act in separate directions. *Indra* X-force (FipaX~FippX) gets separated from Sun X-force (FspaX~FsppX) during the time Planet orbits between aphelion-equinox, equinox-perihelion, perihelion-equinox, and equinox-aphelion.

(iii) *Indra* Z-force (FipaZ~FippZ) at Planet equinox acts at a vertical angle of ½(A+P) upon Planet magnetic field-flow, which creates a low drag force on the Planet and causes the ZZ-axis (at N-pole) tilt up (or down). Sun X-force acts on the Planet at a variable angle, Pi=Pi-0-Pi, to XY-plane, the plane of orbit, as the orbit-inclination, Pi, at aphelion or perihelion decreases to 0 at equinoxes; thus, Sun X-force creates a Z-force (FspaZ~FsppZ) that acts on the Planet at 90 degrees to XY-plane. And, Sun Z-force rotates in XY-plane from perihelion to aphelion side and aphelion to perihelion side as the Planet orbits (as in the case of Earth once a year); this Z-force causes ZZ to spin anti-clockwise and tilt up (or down).

(iv) *Indra* X-force (FipaX~FippX) acting on the Planet in XY-plane (in X-direction, right to left) makes Planet ZZ-axis to spin clockwise at all times… Also, *Indra* X-force makes Earth to rotate about ZZ-axis once in each orbit around the Sun when Earth magnetic-S faces the Sun at Ai=90E (like our-Moon rotating once in each orbit around Earth per NASA).

(v) *Indra* X, Z-forces and Sun X, Y-forces act on the Planet in 2 planes-XZ, XY at 90 degrees to one another; and together make the Planet orbit in XY-plane at an orbit-inclination, Pi (XZ-plane, the plane Sun orbits). To visualize the wobble that comes in 2-small continuous-circles, let the Planet start orbiting from (bottom) equinox (B) towards perihelion:

At equinox (B) since Pi=0, Sun Z-force=0

Indra Z-force=½(FipaZ+FippZ) acts at angle ½(A+P) to XY-plane and tilts ZZ (N-pole) up, y1; *Indra* X-force spins ZZ clockwise, x1: Y=y1, X=x1

At ½ Rpp since Pi=½Pi, Sun Z-force=½(Z-force) tilts ZZ down, -y2; *Indra* Z-force=½(Z-force of equinox) tilts ZZ up, ½y1: so, Y=(1½y1-y2). *Indra* X-force spins ZZ to x2: X=(x1+x2). Y=(1½y1-y2), X=(x1+x2)

At perihelion since Pi=Pi, Sun Z-force=Z-force tilts ZZ down, -y3 plus anti-clockwise spin –xp; vertical angle P=0: so, *Indra* Z-force=0; *Indra* X-force spins ZZ to x3: Y=(1½y1-y2-y3), X=(x1+x2+x3-xp)

To go from perihelion to ½ Rpp (top) and to equinox (T) derive the values similarly as above…

That gives a set of values for one-circle; and going from equinox (T) to ½Rpa (top) to aphelion and return to equinox (B) is the second set for another circle…

One set of X, Y values for the wobble would look like this:

(B) Equinox	½Rpp	perihelion	½Rpp	(T) Equinox
Y y1	(1½y1-y2)	(1½y1-y2-y3)	(1½y1-y2-y3+y2-y3+y2-½y1)	(1½y1-y2-y3+y2-½y1-y1)
X x1	(x1+x2)	(x1+x2+x3-xp)	(x1+x2+x3-xp+x4)	(x1+x2+x3-xp+x4+x5)

The two-sets of Y, X represent the path Planet ZZ-axis at its N-pole takes in each orbit around the Sun; IERS measures the movements of Earth ZZ (at N-pole) and calls it the wobble. The x1 to x5… represent a continuous clockwise spin of Planet ZZ-axis, and y1 to -y3… represent tilt of ZZ-axis due to the 'drag' of Sun Z-force and *Indra* Z-force. Here, x5… is the total spin of ZZ-axis in ½ -orbit.

Note: See Endnotes at *[16] to determine Wobble by others.

<u>Summary of Eq. 2D</u>:

Orbit velocity,

$Vo = \frac{1}{2}\{[(Bspp)(Sm/Rpp)]^{\wedge}\frac{1}{2} + [(Bspa)(Sm/Rpa)]^{\wedge}\frac{1}{2}$
$\quad + [(Bipp)(Im/Rspp)]^{\wedge}\frac{1}{2} - [(Bipa)(Im/Rspa)]^{\wedge}\frac{1}{2}\}$

Sun Y-force that rotates Planet, $FsY = \frac{1}{2}[FspaY + FsppY]C$, In Eq.2a

*C=Constant of interaction between Sun-Planet magnetic fields-intensity—assume
C=1 to 0.9, at Ai=0N-180S, since ZZ-axis stays closer to <u>high</u> magnetic flow; and
C=0 at Ai=0N, Ai=90E, Ai=180S & Ai=270W, since the interaction will be none at these locations.

<u>Note</u>: FsY acts in XY-plane on Planet-diameter (ZZ). The calculations are given at *[16]

Sun X-force creates Z-force (FsZ) at aphelion & perihelion of the Planet due to orbit-inclination, Pi; that FsZ and the Planet magnetic flow, together, creates a <u>high</u> or <u>low</u> 'drag force' on the Planet causing ZZ-axis to wobble and spin anti-clockwise. For the Earth, we deduced the drag force constant, C1, by trial-error using IERS Zs23.5~0.20arcsec/Yr. *See* Table 6 and at *[16]

So, at perihelion Sun's $FsZpp = [\frac{1}{2}(FspaX + FsppX)(Sin\ Pi)][(64/365)C1*]$
At aphelion Sun's $FsZpa = [\frac{1}{2}(FspaX + FsppX)(Sin\ Pi)][(64/365)C2*]$

*<u>High</u>-drag force constant, $C1 = (C)(Sin\ Ai)(Sin\ Ai \times Cos\ Ai)^{\wedge}2$, Ai=0-45, and C=0.9
<u>Low</u>-drag force constant, $C2 = (0.25C)(Sin\ Ai)(Sin\ Ai \times Cos\ Ai)^{\wedge}2$, Ai=0-45

<u>Note</u>: Sun's FsZ at equinoxes is Zero since Pi=0. FsZ acts along Z at 90 to XY-plane, on Planet-diameter (XX) and rotates in XY-plane causing wobble plus anti-clockwise spin of ZZ. The net-effect of Sun's FsZ at Earth aphelion & perihelion estimated to be 64 days out of 365 days (each orbit).See at *[16]

Indra Z-force creates FizY at equinoxes of Planet and causes tilt of ZZ (up or down) to wobble, FizY= ½(FipaZ+FippZ)(Sin ½(A+P))

Note: FizY becomes 0 at aphelion and perihelion; so, the tilt of ZZ varies from 0 to Max. at equinoxes. Thus, Sun's FsZ and *Indra*'s FizY cause Planet ZZ-axis move up and/or down in YY, and a part of the wobble, while *Indra* X-force (or its drag force) causes Planet ZZ spin clockwise. See Eqs., below:

Indra X-force spins ZZ clockwise (Ai=180S-225W or 0N-45E), FiX1=[½(FipaX+FippX)]C1*

*C1=(0.9)(Sin Ai)(Sin Ai x Cos Ai)^2, Ai=0-45;
 C1=<u>High</u> drag-force constant is deduced from estimated IERS Zs23.5~0.20arcsec/Yr. *See at* *[16]

Indra X-force spins ZZ clockwise (Ai=225W-270W or 45E-90E), FiX1=[½(FipaX+FippX)]C1*

*C1=(0.9)(Sin Ai)(Sin Ai x Cos Ai)^2, Ai=45;
 C1=<u>High</u> drag-force constant- as ZZ will be at <u>high</u> flow-intensity of magnetic-N beyond Ai=45 (Ai=45-90 or 225W~270W), the value for C1 remains the same as at Ai=45.

Indra X-force spins ZZ clockwise (Ai=90E-135E), FiX2=[½(FipaX+FippX)]C2*

*C2=(0.25x0.9)(Sin Ai)(Sin Ai x Cos Ai)^2, Ai=0-45;
 C2=<u>Low</u> drag force constant

Indra X-force spins ZZ clockwise (Ai=135E-167E), FiX2=[½(FipaX+FippX)]C2*

*C2=(0.25x0.9)(Sin Ai)(Sin Ai x Cos Ai)^2, Ai=45-77;
 C2=<u>Low</u> drag force constant

Indra X-force spins ZZ clockwise (Ai=167E-180S), FiX2=[½(FipaX+FippX)]C2*

*C2=(0.25x0.9)(Sin Ai)(Sin Ai x Cos Ai)^2, Ai=45-77;
 C2=<u>Low</u> drag force constant- as ZZ will be at <u>high</u> flow-intensity of magnetic-N beyond Ai=77 (Ai=77-90 or 267E~180S), the value for C1 remains the same as at Ai=77.

Note: FiX1-FiX2 act in XY-plane on Planet-diameter (YY) causing clockwise spin ZZ

Proof
Verify: Neutron **mass (Im), and Neutron-Sun distance (Rs)**

It is clear from Eqs.1 & 2c, Newton Gp is ~equal to Bp, the force constant of Sun (+Neutron) on a Planet. The results obtained from the Eqs., using appropriate data, are listed in Table 2. It shows, Bp is ~equal to Newton Gp, for Planet Mercury to Neptune.

Therefore, in Eq.2c the use of Neutron mass, Im ~4×10^{30} kg., and Neutron-Sun distance, Rs ~161×10^{11}m yields an accurate result. We will not get that result or be able to verify the observed NASA data if we were to use any other combination of mass, Im, and distance, Rs. In fact, the use of Chadwick's mass for the neutron in atom H, [1.0067 x mass of (proton + electron)] did not verify the observed NASA data shown in Table 1. However, the use of mass for the Neutron in Atom He [1.006 x 2 (mass of Sun + Planets)] did verify the observed NASA data*[16].

Moon orbits Planet with its magnetic-S or N*[16.2] pointing towards Planet equator. Sun has no direct-control upon Moon orbit. Earth-Neutron controls Moon's orbit-rotation. That must be true of all satellites going around the Earth, including the living-objects on the Earth- otherwise, the Moon, satellites (space crafts)*[15] in space, and the rest of living-objects on the Earth would be pulled away into the hot-Sun.

Furthermore, take Earth orbit data (from NASA) plot a closed oblique-curve for the orbit (at Pi=0.00005 degrees) to a large scale (not solved here); and measure Ei (tangent angle to the curve) and A, P at several points on the curve (Rpa-0-Rpp, then Rpp-0-Rpa). From the plot determine an (approx.) Eq. to describe the curve and verify if it is an oblique-ellipse or some other curve.

Conscious mind of man provides the **means** (ability to act with force) for the man made to move in space or orbit around a Planet at varied-speeds and distances from the Planet. In Macro-structure, Neutron's conscious mind provides the **means** (an ability to act with force) for the Sun-Planets to move at varied-speeds and distances in their orbits. **It is that Reality the man must recognize**.

Experiments

Speed of Radiation v. Light

Light wave is a product of hot-cold dark matter in space; and without Suns' radiation, light cannot appear or speed on its own. Only our ability to perceive the difference between radiation and light is the issue because there is no light without radiation. A sharp eye can sense the light much faster than skin can sense the radiation. Even a firefly cannot display its light-colors (yellow, green or red…) in the absence of making a low-level radiation through a chemical action. At night, the eye can sense that light (not in day light) from a short distance, but our-skin cannot sense that radiation (unless skin comes in contact…).

A simple experiment in a dark room can be set up to measure the speed of radiation. Keep the room at ~280K and send a laser pulse of ~700 nm from one end of the room to a sensor at the other end of the room (a distance of ~20m or so); and record the wavelength of the arrived pulse and the time of its arrival. Then, keep the room at ~45K, and repeat the test and record the data. Repeat the experiment varying the room temperature, wavelength of the pulse and the distance etc…Use the data obtained- such as, the time of flight, loss/gain in wavelength of the pulse- and determine the speed of radiation. Add the speed of radiation and the speed of Suns' hot-matter near the Earth, and get the true speed of light on the Earth. The same experiment can, also, be repeated using hot-water at 300-400K (in a container of 5-10m long…) and an ice-water (in a container of 5-10m long).

Measure distance between Sun-Star (Proxima)

It appears we found the companion-Sun, but unable to identify it as such, precisely- for lack of a reliable method to measure distances in space. Per an engineer-surveyor, a reliable method of measuring distance to an inaccessible point or a Star (Proxima) in our case is the 'triangulation-method'. Simply put- knowing the distance between two points on a base line (Sun-Earth line or Sun-Jupiter line etc..), measure the angle between the base line and the point in space (Star), whose distance is required, from two (or more) points on the base line. Measure at least two angles, one at each end of the base line. In our case, the points on the base line can be- aphelion, perihelion and/or equinoxes.

Obviously, Sun-Saturn or Sun-Jupiter line is preferable over Sun-Earth line to obtain an accurate measure for the angles... Knowing the distance (AU) between these two points on the base line and the angles to an inaccessible point in space, we can calculate an accurate distance between our-Sun and the companion (Star). NASA's Cassini...at Saturn or other missions making various measurements in space... could do this simple survey.

Conclusions:

1. The deduced 'structure of our-universe and the motion of matter' confirms that the varied orbit-distances of the Planets around the Sun and the varied orbit-inclinations to the elliptic (the plane of Sun-Earth line) is the cause of *Indra* (Neutron) at the center of our-universe or Atom.

i. *Indra* is located in a cold-dark space at an average-distance of 161×10^{11}m (108 AU) from our-Sun with a mass of 4×10^{30} kg.

ii. *Indra* makes cold, and orbits the cold matter in XZ-plane in the direction *Indra* rotates.

iii. The orbiting cold matter makes Sun to orbit in XZ-plane and rotate about ZZ-axis in XY-plane.

iv. Sun makes radiation, and orbits the hot matter in XY-plane in the direction Sun rotates.

v. The orbit-speed of the cold matter decreases with increase of distance from *Indra*; so is the orbit-speed of hot matter which decreases with increase of distance from Sun- due to the collisions and interferences between the orbiting cold and hot matter.

vi. The orbiting hot matter and the orbiting cold matter, together, make Planets orbit at varied-distances from the Sun in XY-plane (the plane of Sun-Earth line) and at varied orbit-inclinations (to XY-plane) with aphelion leading, and axis of rotation ZZ spinning toward the Sun-*Indra*. *Indra* X-force makes Planet ZZ-axis spin clockwise (Earths' mantle + crust ~342km thick spins).

(a) Earths' ZZ-axis (at Ai~ 23.5E) spins at a speed of ~0.20arcsec/ year or per orbit; the spin and rotation will become Zero (at Ai=90E). It takes 1.296×10^6 years for Earths' magnetic-S (N-pole) to spin from Ai=0N to 90E and face the Sun (like Uranus). A new *Yuga* begins at Ai=90E (after thousands of years) with Earths' magnetic-S reversing into magnetic-N, a pole reversal. Earth continues to spin and the magnetic-N faces *Indra* at Ai=180S (like Venus); there, again, the spin and rotation will become Zero. It takes ~3.024×10^6 years for Earths' magnetic-N (S-pole) to spin from Ai=90E to 180S. At Ai=180S (after thousands of years) Earths' magnetic-N reverses into magnetic-S and continues to spin (Ai=180S-270W). *Indra* X and Sun Z forces cause the spin of ZZ, and *Indra* X, Z and Sun Z forces cause the wobble of ZZ-axis.

vii. Planets' rotation makes hot and cold matter mix, and orbit the mix in the direction Planet rotates; the orbit-speed of the mix decreases with increase of distance from the Planet due to collisions and interferences between hot and cold matter.

viii. The hot-cold matter mix that orbits (in XY-plane) under Planet rotational-force, and the cold matter that orbits (in XZ-plane) under *Indra* force, together, make Moon to orbit at varied-distances from the Planet in XY-plane in the direction Planet rotates; and Moons' magnetic-S spins toward the Planet, like Earths' magnetic-S spins toward the Sun...

ix. *Indra* Z-force makes Moons to orbit at varied-inclinations to XY-plane, and causes retrograde (or polar) orbits to the outer Moons.

x. *Indra* force controls Suns, Planets, Moons, ring-objects and the rest in our-universe or Atom.

2. Sun has a companion, and the Sun-Star orbit within and around *Indra* in the same direction at different speeds and radii.

i. Orbit-period of Sun is 684-years at an average distance of 108 AU from *Indra*.

ii. Orbit-period of Star is 2500-years at an average distance of 238 AU from *Indra*.

iii. The distance-ratio: distance from (*Indra*-Star)/(Sun-*Indra*) is 2.2 in our-Atom; and the distance-ratio: distance from (neutron-proton2)/(proton1-neutron) is 2.2 in atom-He.

iv. The mass-ratio: (*Indra*)/(Sun+Star) in our Atom is ~equal to (neutron)/(2-protons) in atom He.

3. Sun orbits along equator of *Indra* with Suns' magnetic-N (or S) pointing toward *Indra*, with a pole-reversal every 10-11 years. See below the force-constants of *Indra*, Sun & Star:

i. When Sun & Star are in <u>opposite-side</u> of the *Indra*-
Force-constant at *Indra*, Bii=13.694x10^-11 N.m^2/kg^2,
Force-constant of *Indra* at Suns, Bis~Bics= 7.144x10^-11 N.m^2/kg^2
Force-constant of Sun for Planets, Bss= 6.670x10^-11
Force-constant ratio: Sun/ *Indra*, Bis/Bii ~ 0.52

ii. When Sun & Star are on the <u>same-side</u> of the *Indra*-
Force-constant at Sun, Bisc=10.718x10^-11 (or >1.5Bis).

iii. *Indra* force, being high at its center, it <u>decreases</u> ~linearly between *Indra*-Sun as the radiation, T, in space or the distance from *Indra* <u>increases</u>; *Indra* force continues to <u>decrease</u> beyond the Sun to a point at ~225 AU from the center of *Indra* or ~117 AU from the Sun.

iv. *Indra* force, being high at its center, it <u>decreases</u> ~linearly between *Indra*-Star as the radiation, T, in space or the distance from *Indra* <u>increases</u>; *Indra* force continues to <u>decrease</u> beyond the Star to a point at ~497 AU from center of *Indra*.

v. Our-universe, an Atom (*Anuvu*), is an elongated-sphere of variable diameter ~722 AU.

vi. *Indra* force (through Sun) on the Planet <u>increases</u> ~linearly (at a flat-slope) as the radiation, T, in space or the distance from *Indra* <u>decreases</u>.

vii. Sun has <u>no</u> direct-control upon Moons' orbit; Earth and *Indra* controls the Moon. That must be true of all satellites orbiting around the Earth, and other objects on the Earth; and it must be true of all the Moons (ring-objects) at the other Planets.

4. Sun and Star will be on the same-side of the *Indra* at ~130 AU apart, once in 942 years.

i. Sun being in its last ¼ lap of the present-cycle around the *Indra*, it will approach the Star on the same-side of *Indra* with the distance between them decreasing from ~216 to 130AU. Then, the distance between Sun and Star increases from ~130 to 216AU in the first ¼ lap of a new cycle.

    ii. During the next ~200 years, Earth and other Planets orbit closer to the Sun as the force-constant of the Sun increases by ~50% (Bisc >1.5Bis). That could make the temperature on Earth to rise, and melt polar-ice at a faster phase than at the present.

5. Use the Planet force-constants at perihelion, Bpp, and at aphelion, Bpa, for the combined force of *Indra* + Sun to calculate the variable surface gravity on the Planet at any point between perihelion-aphelion (Rpp, Rpa) or any point in space near the Planet.

    i. The variable-surface gravity defines dynamic-weather on the Planet (Moon or Sun).

6. The cold-Magnetic force of the dark matter in space <u>increases</u> with <u>decrease</u> of radiation, T, and hot-Electric force of the dark matter in space <u>increases</u> with <u>increase</u> of radiation, T.

    i. Sun inhales cold-hot dark matter and Earth (other living) inhales cold-warm dark matter, and converts that into hot-Electric force and exhales.

7. Like our-universe, all other Atoms in our-galaxy and all galaxies must be bound together by *Indra*s, and perform a duty.

    i. The distance-ratio: (distance from *Indra*-Sun)/(distance from Sun-Earth) is 108AU/1AU~108. And,

    ii. The force-constant ratio: (Sun)/(*Indra*) is Bis/Bii~ 0.52. These 2-ratios must be the Universal-constants in our-galaxy.

8. Suns' rotation-speed plus sunspot pressure makes the hot-matter to orbit at a radial-speed of ~436,750 m/s (at Suns' surface), and the hot matter forms into hot-Electric field-flows. From Suns' equator the cold-hot matter flows toward magnetic-S; and from Suns' magnetic-N the cold-hot matter flows toward the equator and the hot matter flows toward *Indra*. Similar flow pattern must be true at the Planets, Moons and other objects in our-universe.

i. *Indra*s' rotation-speed makes the cold-matter to orbit at a radial-speed of ~97,700 m/s (at ~68.85 AU from *Indra* or ~39.15 AU from the Sun); the cold-matter must form into cold-Magnetic field-flow from its equator to the magnetic-S, and from its magnetic-N to the equator.

ii. *Indra* controls Suns' orbit, rotation and tilt. *Indra*s' cold-Magnetic field-flow creates a trigger at ~39.15 AU from the Sun, and flips Sun poles (magnetic-N to S or S to N) every 10-11 years.

iii. *Indra* force in space or on the Planet increases as the distance from *Indra* decreases; that increase in force causes an increase in the wind-speed (Ws) on the Planet.

iv. The wind-speed (Ws) on a Planet is directly proportional to:

(a) The orbit-velocity (Vip) of the Planet under *Indra* force, and

(b) The surface-speed (Ss) of the Planet under Sun force.

When the speed of hot matter increases abruptly, the hot matter pulls the cold matter into spiral circles as a tornado on the Earth (or Sun) and as a cyclone in Ocean.

9. The dark matter in space disperses (absorbs-emits) Sun radiation, T, and appears in the form of electro-magnetic waves. The speed of the wave decreases between Sun-*Indra* as the distance from Sun increases or the radiation, T, in space decreases.

i. The Cosmic Microwave Background (CMB), T ~2.7K, observed in space, at stations located near Earth, must be:

(a) The left-over radiation of distant-Stars at distant-*Indra*s, and/or

(b) The left-over radiation of distant-Stars plus the left-over radiation of Sun at *Indra* (including that coming from the collisions at <u>nose</u> and <u>magnetic-N</u> with *Indra*s' cold matter).

ii. The speed of light in space is not a Universal constant; the speed varies and it depends directly upon radiation, T, in space or entropy gradient, and the force-constants at the point of interest.

iii. The speed of radiation coming from a distant-emitter to the Earth varies from point to point in its path due to the presence of *Indra*s and cold-dark matter- speed, c (~3x10^8m/s), relate to T 280-295K.

10. The cold-Magnetic force of the dark matter in space <u>increases</u> as the radiation, T, <u>decreases</u>; and the hot-Electric force of the dark matter in space <u>increases</u> as the radiation, T, <u>increases</u>.

i. The cold-Magnetic and hot-Electric force in space form into a 'band' and acts as a 'tension' and/or 'compression' spring between *Indra*-Sun, Sun-Planets, Planet-Moons, *Indra*-Planets & *Indra*-Moons…and keeps them in proper motions.

Tables:

Table-1: Physical Characteristics-Orbits of Planets (See NASA's Solar System Exploration)

	Mass(kg.)	Diameter(m)	Sun's Distance (m) Rpa/Rpp (Pi or Mi)	Mean Orbit-velocity (m/s)
Sun	1.989×10^{30}	1.391×10^{9}	———	———
Mercury	3.300×10^{23}	0.488×10^{7}	0.6982×10^{11} 0.4600×10^{11} (7)	47872
Venus	4.869×10^{24}	1.210×10^{7}	1.08942×10^{11} 1.07476×10^{11} (3.39)	35021
Earth	5.974×10^{24}	1.276×10^{7}	1.521×10^{11} $1.471 \times 10^{11} (0.5 \times 10^{-4})$	29786
Earths' Moon	0.73483×10^{23}—		4.055×10^{8} 3.633×10^{8} (5.145)	1023
Mars	6.419×10^{23}	0.679×10^{7}	2.492×10^{11} 2.066×10^{11} (1.8)	24131
Jupiter	1.899×10^{27}	1.430×10^{8}	8.160814×10^{11} $7.407426 \times 10^{11} (1.305)$	13070
Saturn	5.685×10^{26}	1.210×10^{8}	15.03983×10^{11} 13.49467×10^{11} (2.484)	9672
Uranus	8.685×10^{25}	0.510×10^{8}	30.0639×10^{11} 27.3556×10^{11} (0.77)	6835
Neptune	1.024×10^{26}	0.500×10^{8}	45.3687×10^{11} 44.5963×10^{11} (1.769)	5478
Pluto	13×10^{21}	0.1151×10^{7}	44.3682×10^{11} (17.14) 73.7593×10^{11}	4749

Table-2: Planet & Sun Orbit: Results for 'force constants' Bpp&Bpa, **Bp**, Bsp & Bip

	Mass (kg) (10^30)	Mean Distance (Rs) (10^11m)	Force constants Bss/Bsi <= (10^-11)	Bis/Bii =>
Neutron	**4.00**			**7.144/13.694**
Sun	1.989/1.997	**161.00**	**6.670/7.144**	

Orbit-

	Newton G_p (10^-11)	Velocity Vpp^2&Vpa^2 (10^8m/s)	Bpp/Bpa (**Bp**) <=	Bspp/Bspa (Bsp+/-Bip) (10^-11)	Bipp/-Bipa) (Bip) =>
Mercury	6.670688	28.844824/ 19.009483	*6.6669*	6.69087/ 6.64287	6.6718	0.01907 -0.02893
Venus	6.672075	12.347049/ 12.181575	*6.6716*	6.71648/ 6.62679	6.6719	0.04454 -0.04515
Earth	6.673000	9.022840/ 8.726231	*6.6717*	6.73396/ 6.60940	6.6730	0.06095 -0.06362
Moon	6.734000	0.011073/ 0.009921	*6.7341*	6.73429/ 6.73397	6.7341	0.00015 -0.00017
Mars	6.671511	6.423340/ 5.325284	*6.6634*	6.75787/ 6.56900	6.6723	0.08560 -0.10326
Jupiter	6.685377	1.795017/ 1.629309	*6.6710*	6.99387/ 6.34804	6.6865	0.30732 -0.33851
Saturn	6.671024	0.988973/ 0.887380	*6.6840*	7.27681/ 6.09123	6.7159	0.56093 -0.62464
Uranus	6.743287	0.490277/ 0.446111	*6.7085*	7.89528/ 5.52175	6.7625	1.13278 -1.24133
Neptune	6.786610	0.302701/ 0.2975475	*6.7879*	8.61838/ 4.95747	6.8025	1.81588 -1.84504
Pluto	6.697000	0.3002225/ 0.1805919	*6.6468*	9.09882/ *4.19470	7.1889	1.91000 -2.99420

*Effect of Neutron's trigger-line   Bpp=Bsp+Bipp, Bpa=Bsp-Bipa
Bp=(Bpp+Bpa)/2

Table 3:

	Mercury	Venus	Earth	Mars	Jupiter	Saturn	Uranus	Neptune	Pluto
Rp(10^11)	0.5791	1.0821	1.496	2.279	7.7841	14.2673	28.71	44.983	58.6
Vsp(m/s)*1	47,786	34,904	29,648	23,962	12,765	9,270	6,300	4,861	4,059
Vip(m/s)*2	86	117	138	169	305	402	535	617	690
Bhp(10^-11)*3	6.648	6.628	6.611	6.579	6.377	6.164	5.729	5.344	4.854
K*4(5777)	671	381	280	187	56	31	15	10	7.5

*1 Planet orbit-speed in XY-plane caused by Suns' force decreases beyond Neptune at ~60m/s per 10^11m; and Planets' orbit and hot-matter's orbit are in the same direction (see Eq.5f).
*2 Planet mean orbit-speed in XY-plane under Neutron force (due to orbit-inclination) increases beyond Neptune at ~5.5m/s per 10^11m; and Sun (taking Planets with it…) and cold-matter orbit in the same direction (see Eq.5i).
*3 Bhp on the Planet under Sun hot-matter decreases beyond Neptune at 0.03x10^-11 per 10^11m
*4 See Table 4 & 4a

Table-4: Distance from the source v. Radiation absorb-emit (or disperse) in space, XZ-plane

*K per NASA (or at corona 1x10^6K)

Name	Distance (10^11m)	K	X, m (10^-6)	Gravity (G) m/s^2
Sun	0.00696	*5777.0(1x10^6K)	0.0266(0.000165)	274.26
—-	0.58	671.0	0.2290	—
—-	1.08	381.0(~409K)	0.4034(0.403741)	—
—-	1.496	280.0(~295K)	0.5500(0.560200)	—
—-	2.28	187.0	0.8220	—
—-	7.78	56.0(~56.5K)	2.7410(2.923645)	—
—-	14.27	31.0	5.0060	—
—-	28.71	15.0	10.0400	—
—-	44.98	10.0	15.720	—
——	59.06	7.50(~7.5K)	20.630 (22.21015)	—
Neutron	161.00	*2.75(~2.7K)	55.850(60.55000)	0.0974

For Sun Xs=0.34891(0.376105)x10^-6m/10^11m

For c-Sun Xcs ~0.15681x10^-6m/10^11m

<u>Table</u>-4a: Distance from the source v. Radiation absorb-emit (or disperse) in space, XY-plane

*K per NASA (or at corona 1x10^6K)

Name m/s^2	Distance (10^11m)	K	X, m (10^-6)	Gravity (G)
Sun	0.00696	*5777.0(1x10^6K)	0.0266(0.000165)	274.26
Mercury	0.58	671.0	0.2290	3.70
Venus	1.08	381.0(~409K)	0.4034(0.403741)	8.88
Earth	1.496	280.0(~295K)	**0.5500(0.560200)	9.80***
Mars	2.28	187.0	0.8220	3.72
Jupiter	7.78	56.0(~56.5K)	2.7410(2.923645)	24.85
Saturn	14.27	31.0	5.0060	10.43
Uranus	28.71	15.0	10.0400	9.03
Neptune	44.98	10.0	15.7200	11.15
Pluto	59.06	7.50(~7.5K)	20.6300(22.21015)	0.69
——	161.00	*2.75(~2.7K)	55.8500(60.55000)	0.0974

**For Sun per Young (1802) visible light, X=0.0000203"(27.07/1000)=0.55x10^-6m

***Earths' gravity at Sun=Neutron=0

<u>Table</u> 5: (From Eq.6.3)

Planet	Ws per NASA	Ss	Vip*3	Ws
Mercury*1		11 km/hr.	310 km/hr.	**1 km/hr.**
Venus		-6.5	421	**~1**
Earth	240km/hr.	1,670	497	**240**
Mars		867	608.5	**152**
Jupiter	600	45,260	1098	**600**
Saturn	1,800	35,535	1447	**1800**
Uranus		(0+9,300+)/2*	1926	**-770+**
Neptune*2		9,660	2221	**3080**

*1 Vip mix at Mercury will have mostly hot-matter;

*2 Vip mix at Neptune will have mostly cold-matter
*3 See Table 3 for Vip m/s converted into km/hr.
*Uranus surface speed, -Ss=0 when its N-S poles point toward Sun (or Ai=90d) at solstice; and >9,300+km/hr. when its equator point toward Sun at equinox…(so, NASA data of 9,300km/hr. must be erroneous).

Table-6: IERS data (EOP C04) on Earth polar motion at Solstice: X, Y (arc sec)

Ra or $Rp = [(X^2)+(Y^2)]^{0.5}$ arc sec

Date	Aphelion (June-July?)			Perihelion (Dec.-Jan.?)			Wobble?	
	X	Y	*Ra*	X	Y	*Rp*	D~ *Ra+Rp*	*(Ra-Rp)*
1973 2 15				40.112	84.612	*0.0936*		
1973 9 22	29.272	330.272	*0.3316*				*0.4252**	0.238
1974 3 15				53.220	165.120	*0.1735*		
1975 3 25	27.223	340.023	*0.3411*				*0.5146*	0.168
1975 10 25				2.681	94.381	*0.0944*		
1976 5 10	4.335	435.335	*0.4354*				*0.5298*	0.341
1976 12 13				1.995	73.095	*0.0731*		
1977 6 21	38.347	487.447	*0.4890*				*0.5621*	0.416
1978 1 1				0.400	25.300	*0.0253*		
1978 7 25	40.956	487.856	*0.4896*				*0.5149*	0.464
1979 2 15				7.912	58.812	*0.0593*		
1979 9 9	47.769	410.569	*0.4133*				*0.4726*	0.354
1980 3 10				48.719	178.819	*0.1853*		
1980 12 9	42.794	374.094	*0.3765*				*0.5618*	0.191
1981 8 20				12.163	183.763	*0.1842*		
1982 3 6	41.517	443.817	*0.4458*				*0.6300*	0.262
1982 9 30				32.774	68.374	*0.0759*		
1983 4 22	23.750	560.910	*0.5615*				*0.6374*	0.485
1983 11 12				42.286	15.016	*0.0449*		
1984 6 21	74.398	557.822	*0.5628*				*0.6077*	0.518
1985 1 2				38.735	22.066	*0.0446*		
1985 8 2	56.926	496.130	*0.4994*				*0.5440*	0.455

Date	Aphelion (June-July?)			Perihelion (Dec.-Jan.?)			Wobble?	
	X	Y	*Ra*	X	Y	*Rp*	D~ *Ra+Rp*	(*Ra-Rp*)
1986 4 2				-45.029	136.257	*0.1435*		
1986 10 15	76.549	390.959	*0.3984*				*0.5419*	0.255
1987 6 15				11.890	198.003	*0.1984*		
1988 2 14	54.165	435.419	*0.4388*				*0.6372*	0.240
1988 9 30				13.242	129.642	*0.1303*		
1989 4 24	94.286	481.571	*0.4907*				*0.6210*	0.360
1989 11 14				1.654	75.494	*0.0755*		
1990 6 5	59.882	575.914	*0.5790**				*0.6545*	0.503*
1990 12 26				47.946	69.333	*0.0843**		
1991 7 24	64.191	573.141	*0.5767*				*0.6610*	0.492
1992 3 1				9.217	114.950	*0.1153*		
1992 10 5	65.432	502.893	*0.5071*				*0.6224*	0.392
1993 4 29				41.686	164.343	*0.1695*		
1994 1 8	29.092	480.143	*0.4810*				*0.6505*	0.312
1994 8 20				43.455	168.714	*0.1742*		
1995 4 11	62.691	558.202	*0.5617*				*0.7359*	0.387
1995 10 30				31.671	81.862	*0.0878*		
1996 5 30	57.654	596.950	*0.5997*				*0.6875*	0.512
1996 12 14				36.526	85.288	*0.0928*		
1997 7 4	31.305	537.062	*0.5380*				*0.6308*	0.445
1998 1 27				25.666	166.942	*0.1689*		
1998 8 15	35.514	478.505	*0.4798*				*0.6487*	0.311
1999 3 18				49.806	238.846	*0.2440*		
1999 12 30	42.493	378.512	*0.3809*				*0.6249*	0.137
2000 8 31				50.133	241.879	*0.2470*		
2001 3 28	81.755	491.313	*0.4981*				*0.7451*	0.251
2001 10 12				27.487	117.390	*0.1206*		
2002 4 22	34.884	555.152	*0.5562*				*0.6768*	0.436
2002 12 2				0.989	141.410	*0.1414*		

Date	Aphelion (June-July?)			Perihelion (Dec.-Jan.?)			Wobble?	
	X	Y	*Ra*	X	Y	*Rp*	D~ *Ra+Rp*	*(Ra-Rp)*
2003 6 11	51.780	548.398	*0.5508*				*0.6922*	0.409
2004 1 2				28.865	153.841	*0.1565*		
2004 7 18	35.713	521.020	*0.5222*				*0.6784*	0.366
2005 2 13				59.241	203.203	*0.2116*		
2005 8 18	27.134	429.401	*0.4303*				*0.6419*	0.219
2006 9 28				39.851	252.190	*0.2553*		
2007 4 30	77.167	491.128	*0.4971*				*0.7524*	0.242
2007 10 26				61.621	190.035	*0.1998*		
2008 5 15	60.248	542.608	*0.5459*				*0.7457**	0.346
2008 12 14				44.769	134.277	*0.1416*		
2009 6 15	82.228	541.830	*0.5480**				*0.6896*	0.406
2009 12 27				108.657	191.751	*0.2204**		
2010 7 7	80.981	484.337	*0.4911*				*0.7115*	0.271
2011 1 25				58.225	195.809	*0.2043*		
2011 7 24	90.995	449.731	*0.4588*				*0.6631*	0.255*
2012 1 10				101.787	257.252	*0.2766*		
End of IERS data…	———	———			———	———		———
Total	14935.	16.0763			4.9138	19.9607	11.163	

Endnotes:

*1 Per NASA, the Earth is immersed in honey and as the Planet rotates, honey around it would swirl and it is the same with space-time. NASA measured geodetic effect, and frame dragging- the amount a spinning Earth pulls space-time with it as it rotates- and claims it determined the effects with precision by pointing at a single Star, while the telescope with gyroscopes made polar orbit around Earth at 642 km above. Polar orbit around Earth meant- similar to the retrograde orbits of outer Moons of Jupiter, Saturn and Neptune....

The result obtained in NASA's experiment (gp-B), if measured correctly, should be identical to the data obtained by International Earth Rotation and Reference Systems Service (IERS) in Paris on Earth's wobble-tilt of its axis of rotation. There is no evidence that NASA has ever compared its result with that of IERS before making a statement that its experiment confirmed two of the most profound predictions of Einstein universe, having far reaching-implications across astrophysics research. In fact, NASA's drift (6.6 arc sec/yr.) and IERS meridian plane wobble-drift, $(Ra+Rp)/2{\sim}0.20$ arc sec/Yr at Ai=23.5E should be identical-because of the <u>cause</u> of the mystery is the 'structure of our-universe'.

The fact defines a fundamental principle of the Nature, not an argument based on complex Eqs. of fiction. Fig.3 [i] shows four Gyro-ellipses of various sizes. Everitt [i] cites "three unforeseen effects emerged.." (due to spin-torque failure), and such failures were "quantified during the calibration phase..". Will [ii] reports "five-years of data analysis were needed to tease out the effects of relativity from...other disturbances..., Everitt... couldn't fully exploit the calibrating effect..., Gyro 2... had the largest uncertainties because it suffered the most resonant jumps."

It is clear the data, or the experiment is erroneous- no amount of fixing would make the four Gyros go correct when the forces made them to get lost. Per Will [ii] the experiment is designed to measure Earth's "warp" and "drag" of the space time. The Earth's warp and drag force or the electromagnetic force, is maximum in its equatorial-plane; the Gyros in a polar orbit encounter + and − forces at Earth equator, and that causes the Gyro "jumps" or the "unforeseen effects". The "tease(d) out" value of 6.6 arc sec/Yr would have been much closer to ~0.20 arc sec/Yr[ix] if the Gyro would take a different-path or not cross Earth equator. A change in Gyro-orbit will provide a useful data on Earth's effect on the Gyro, or Neutron-Sun effect upon the Earth. A simple, quick, and inexpensive experiment, re-using the Gyros, can give us an accurate data if Everitt, Will and NASA would agree.

The Lense-Thirring effect of Earth (on Lageos, Lares, Grace) had a model-error of 60-90% (arXiv:1009.3225 p.9) per Iorio [iii]. In his (1916) erroneous lunar theory that Sun-Earth controls the orbit of the Moon, de Sitter [vi] neglected Earth and Moon orbit eccentricity and orbit-inclination (p.166), and gave support to *1.1Einstein theory on GR, pleading "ignorance" in regard to the origin of inertia by a distant mass (p.183). Chadwick [vii: p.700-2] 1932 hypothesis that "neutron…consists of a proton and an electron in close combination…, find(ing)… the mass of the neutron is 1.0067x mass of (proton + electron)"- an invisible-insensible particle- in hydrogen (H) came too late for de Sitter to learn about origin of inertia.

Such a theory- the experiments, data, and the result- must be erroneous and could keep us going in cycles; therefore, we should be willing to look at the Reality [x], and move forward into the new frontiers of discovery. The Figure 1b shows Neutron-Sun acting upon the Planet, and Neutron-Planet acting upon the Moon. The result verifies the forces from Neutron & Sun control the motion of a Planet (or Neutron & Planet of the Moon….), a Fundamental principle of the Nature. Hence, the curved space-time is a fiction, not the Reality. Once we accepted Chadwick's hypothesis on the existence of a neutron in a living atom, it should be easy to accept the existence of a Neutron in the Atom we live- given the evidence that the motion of matter in the 'structure' verifies all observed NASA data. We appear to be ready to accept 'dark energy' per Boughn [viii], although in Reality only the dark matter (params) exist, not the dark energy; and the published work in PRL[see viii] cannot deduce the size of that dark matter or object, nor the location.

Verlinde[v] concludes [p.9] "origin of gravity.. is an entropic force.. [p.22]. If gravity is emergent, so is space time geometry.., and both have to be given up if we want to understand.. at a more fundamental level... We identified a cause.. for gravity.. driven by differences in entropy.." Based on Verlinde vision, Braunstein[iv] concludes[p.1] "Our analysis is not wedded to GR and.. we find the BH area must be replaced by some other property.. [p.4] in a sense then, black holes are not ideal but 'real black bodies' that satisfy conservation laws, result in a non-thermal spectrum, and preserve thermodynamic entropy..." Although Verlinde, Braunstein and others have put forward the cause as the entropy for Newton gravity or Einstein curved space time, they have not, yet, identified the cause for the 'entropy' itself or why the change in temperature or entropy gradients (or red-shifts)?. Certainly, speaking of BH evaporation as radiation, somewhere in the Universe, is not the answer- 'Neutron' is the cause with a mass, Im, at ~0K, rotating at certain speed, and located at certain distance from our-Sun: see[x].

References: [i]. C.W.F.Everitt, et al., Gravity Probe B: Final Results of a Space Experiment to Test General Relativity, PRL 106, 221101 (2011)

[ii]. C.M.Will, Viewpoint on: Gravity Probe B: Final Results of a Space Experiment to Test GR, Physics 4, 43 (2011)

[iii]. L.Iorio, et al. Phenomenology of the Lense-Thirring effect in the Solar System, Astrophysics and Space Science, Vol. 331, Issue 2, p.351-95 (2010); arXiv:1009.3225

[iv] S.L.Braunstein and M.K.Patra, Black Hole Evaporation Rates without Spacetime, PRL 107, 071302 (2011)

[v] E.P.Verlinde, On the Origin of Gravity and the Laws of Newton, arXiv:1001.0785v1 (or J.High Energy Phys. 04 (2011) 029).

[vi]. W.de Sitter, Einstein's theory of gravitation and its astronomical consequences, MNRAS, Vol. 77, p.155-184 (1916)

[vii]. J.Chadwick, The existence of a Neutron, Proc. Roy. Soc., A, 136, p.692-708(1932)

[viii]. S.Boughn, A distorted view of the early universe, Physics 4, 53(2011) (A Viewpoint on: Das et al., PRL 107, 021301(2011) and Sherwin et al., PRL 107, 021302(2011))

[ix]. IERS data: EOP C04 combined series

[x]. M.R.Mikkilineni, et al., The structure of our-universe and the motion of matter

*1.1 Einstein started with the constant speed of light as a postulate in 1905 in doing special relativity; but by 1911 he was getting into general relativity, and in 1915 said "..that the theory of relativity is still in need of generalization, in the sense that the principle of the constancy of the velocity of light is to be abandoned." Again, in late 1915 he said "the principle of the constancy of the velocity of light in vacuo must be modified" In 1916, he really spelled it out (in a book on Relativity) ".. our result shows that, according to the general theory of relativity, the law of the constancy of the velocity of light in vacuo, which constitutes one of the two fundamental assumptions in the special theory of relativity.., cannot claim any unlimited validity. A curvature of rays of light can only take place when the velocity of propagation of light varies with position. Now we might think that as a consequence of this, the special theory of relativity and with it the whole theory of relativity would be laid in the dust. But in reality this is not the case. We can only conclude that the special theory of relativity cannot claim an unlimited domain of validity; its results hold only so long as we are able to disregard the influences of gravitational fields on the phenomena (e.g. of light)..." And, in his 1920 Leyden address: "According to this theory the metrical qualities of the continuum of space-time differ in the environment of different points of space-time, and are partly conditioned by the matter existing outside of the territory under consideration. This space-time variability of the reciprocal relations of the standards of space and time, or, perhaps, the recognition of the fact that 'empty space' in its physical relation is neither homogeneous nor isotropic, compelling us to describe its state by ten functions (the gravitation potentials gμv), has, I think, finally disposed of the view that space is physically empty".

*1a NASA says- Swift detected new insights into a cosmic accident that has been streaming X-rays towards Earth since late March (2011)- high energy flares from the new source in Draco. It behaves unlike anything we have seen before.. truly extraordinary event- the awakening of a distant galaxy's dormant BH as it shredded and consumed a star.. (and) it took the light from the event approx. 3.9 billion years to reach Earth. The BH may be $2 \times (4 \times 10^6 M)$ or twice the mass of the BH in Milky way. As the star falls towards a BH it is ripped apart by intense tides. The gas is corralled into a disk that swirls around the BH and becomes rapidly heated to temperatures of millions of degrees. The innermost gas in the disk spirals towards the BH, where rapid motions and magnetism create dual, oppositely directed "funnels" through which some particles may escape. Jets driving matter at velocities $>0.9c$ (c=speed of light) form along the BH's spin axis. In the case of Swift, one of these jets happened to point straight at Earth. The radio emission occurs when the out-going jet slams into the interstellar environment. By contrast, the X-rays arise much closer to the BH, likely near the base of the jet. When first detected, the flares initially assumed to signal a gamma-ray burst, one of the nearly daily short blasts of high-energy radiation often associated with the death of a massive star and the birth of a BH in the distant Universe. But as the emission continued to brighten and flare, astronomers realized that the most plausible explanation was the tidal disruption of a sun-like star seen as beamed emission. Zauderer's team showed a brightening radio source centered on a faint galaxy near Swift's position for the X-ray flares. These data provided the first conclusive evidence that the galaxy, the radio source and the Swift event were linked. Observations show that the radio-emitting region is still expanding at $>0.5c$. By tracking this expansion backward in time, we can confirm that the outflow formed at the same time as the Swift X-ray source.

*2 Prof. Krimigis reports that "the radial component of the velocity has been decreasing almost linearly.. and it seems Voyager-1 has entered a finite transition layer of zero-radial-velocity plasma flow- indicating that the spacecraft may be close to the heliopause- the border between the heliosheath and the interstellar plasma. The existence of a flow transition layer in the heliosheath contradicts current predictions". To find more on the reality- see*2.1, 2.2 & 2.3

*2.1 From: Nathan Schwadron [mailto:nschwadron@guero.sr.unh.edu]
Sent: Wednesday, December 21, 2011 8:59 PM
To: Mikki
Subject: Re:

Hi Mikki
My best wishes to you on the holidays. The ribbon is quite unexplained in my opinion. Ideas are welcome.
Nathan

From: Mikki
Sent: Thursday, December 22, 2011 10:25 AM
To: 'Nathan Schwadron'
Cc: 'eberhard.moebius@unh.edu'; 'McComas, Dave'; 'Edward Stone'; 'Krimigis, Tom'
Subject:
12/22/2011

Dear Professors:
I am still reading your 2009 articles in Science… and I am sure I will have many more Qs to ask you for my own understanding of your work…
Why the 'ribbon & knots' appear at this time only in the near-face (or in the view as shown in Fig.3: "IBEX observations…" by McComas et. al. of 09/2011) and not in the far-face side of the 'heliosphere'? I hope I am reading your data and the Fig. correctly- that you discovered the 'ribbon' only in the near-face. If that is true- when

Sun's poles flip in a 11-year cycle- you can expect to see a similar 'ribbon' in the far-face side and <u>not</u> in the near-face... And, my calculations show this 'ribbon-knots' (especially the portion closer to latitude -30d) should be at ~40AU from our-Sun... So, what is this 'ribbon'?

I think, the 'ribbon' is similar to a 'rainbow'...

As you know, 'heliosphere' has to put up with radiated-rays (or light-rays) generated at three locations: heliospheric S pole area, HP area (plus in between), and the Sun. These light-rays arriving from different directions at various intensities converge in 'this area' (where we observe the 'ribbon') and "lit" the "*params*" in space (a *Sanskrit* word or 'photons' as you and Einstein may want to call) to display the "ribbon-knots", a 'rainbow'.

We can devise a simple, inexpensive experiment in a 'dark room' and determine the location of the 'ribbon' in reference to the 3 way laser-light sources... There is more to discover asking a Key Q: What is the cause for the Fire at the S pole, HP & Sun?

We can find a clue from Newton (not Einstein...) to come up with a correct answer to that Q.

I will be happy to work with you and consider it as a privilege and fortunate...

Referring to your "The Interstellar Boundary Explorer Science Operations Center" on p.233 under Fig. 12: The incident fluxes (*left*) and resulting ENA count rates at IBEX (*right*). You said "However, the net compression between the up and downstream plasma over a broad range shows a transition from ~400 km/s to ~100 km/s, and is roughly consistent with our assumed compression." I am not sure what this 400km/s and 100km/s 'up and down' stream plasma speeds refer to? Or at what location with reference to HP or Sun? If we refer to Prof. Krimigis Fig.1b "Zero outward flow velocity...." article in Nature of 06/2011, Vr~100km/s at ~100AU from our-Sun is correct per my own calculations using modified Newton Eq...And, I agree with Prof. Krimigis conclusion of near zero velocity at ~116 AU

Maheswar

*2.2 From: Mikki
Sent: Friday, December 30, 2011 5:25 PM

To: 'Krimigis, Tom'; 'Edward Stone'
Subject: Voyager 1 & 2
12/30/2011

Dear Professors:

In response to my last e-mail request Dr.Krimigis sent me NASA JPL Voyager webpage; and, I spent some time reviewing it and I learned few things…

(1) The speed of Voyager V1 stays at ~17 km/s and Voyager V2 at ~15 km/s

(2) The trajectory of V2 & V1 (as shown in the Fig. below) is in the orbit-direction of the planets around the Sun: that meant in the direction of Sun's rotation…?

(3) The trajectory of V1 shoots-upward (or above the Sun-Earth plane at ~36 degrees), and the trajectory of V2 shoots-downward (or below the Sun-Earth plane at ~48 degrees)….?

(4) V1 encounters 'magnetic clouds' or foamy Fluff… in 2009 & 2011; and V2 did not, yet…?

(5) V1 (at ~119 AU from Sun) & V2 (at ~97 AU) still work, not burnt at the HP despite the high temperature (~10^6 K) that causes the 'ribbon-knots'…?

(6) I did not find the present position of V1 or V2 in terms of X,Y,Z coordinates with reference to Sun (or Earth)- I assume none exists ?

If the above is correct, the Voyagers have no choice but to move with the flow of the 'magnetic cloud'- not against the 'cloud'. As I stated earlier my Calculations show the 'magnetic clouds' move at ~90 km/s towards the Sun, and make the Sun move at ~5 km/s… with ISM flow… Who makes these 'clouds' move that fast? We will get to that, later… Therefore, the 'clouds' that make even the Sun to obey will not look the other way and let the Voyagers swim against the tide [unless Voyagers are powerful to move at speeds >(90+5+17)km/s]. For that reason, I think V1 (at a net speed of 17+5 km/s) and V2 (at 15+5 km/s) have moved towards the tail-end of the 'heliosphere'- not towards the 'nose' or HP where they could have burnt due to high-temperature.

Please give a careful review into this issue and advice… if I am correct or not?

Thanks, Wish you a successful New Year! Maheswar

HELIOCENTRIC VIEW OF TRAJECTORIES

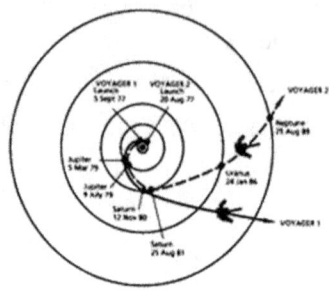

*2.3 From: Mikki
Sent: Sunday, January 08, 2012 9:32 AM
To: 'McComas, Dave'; 'Edward Stone'
Cc: 'Nathan Schwadron'; 'Krimigis, Tom'; 'eberhard.moebius@unh.edu';
'mopher@bu.edu'
Subject: which way Earth (or Sun) orbits ?
01/08/2012

Dear learned Professors:

Nice to write to you all, again, in early in January 2012 with a hope to get certain key issues resolved. Please look into the Qs I raise…. whatever may be the impact…?

In my last e-mail message of 12/30/2011 (to Dr.Krimigis and Dr.Stone), I pointed out why I think Voyager V1 & V2 (at a speed of ~17km/s & ~15km/s) may have reached the 'tail-end' of the 'heliosphere', <u>not</u> the 'heliosheath (HP). The speed of the Inter Seller Medium (ISM) at ~90km/s is one of the reason- the ISM makes Sun move at a speed of ~5km/s in the direction of ISM (not against ISM flow).

I am still reviewing some of your articles, selectively.

(1) In this process on Friday, 01/06/2012, I happened to see Dr.Opher "*A strong, highly tilted* ISM *field near* S*olar* S*ystem*" Nature (2009). It says on p.1037: Sun moves relative to the local Std. of Rest at 13.4km/s

(-9km/s in radial dir., and 3.7km/s in vertical dir.); and Vs=88.09km/s, the shock velocity.

……..

(2) I, also, saw NASA webpage that describes IBEX with the 3-illustrated images, below:

- Looking at image 1- which way is the Earth moving- whether in the direction of the flow of Sun's wind?

- Looking at image 2- which way is the Earth's magnetic flow- directly from N to S pole, or N pole to equator, and equator to S pole? Do we have IBEX data that describes the actual-flow lines?

- Looking at image 1 (assuming the 2-fingers, long & short, seen to the left are the telescopes that pick-up ENA- H, He etc… at 100eV to 6keV…)- what is this energy at 1keV or 1000eV meant? Does that mean the IBEX system generates an energy signal at 1602x10^-19J, and broadcasts that signal into the vacuum thru one of the telescopes and catches 'suitable partners' (like ENAs)?

Thank you, Maheswar

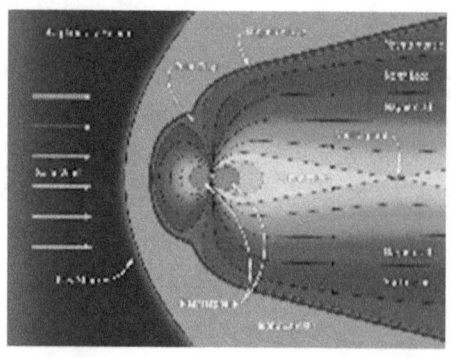

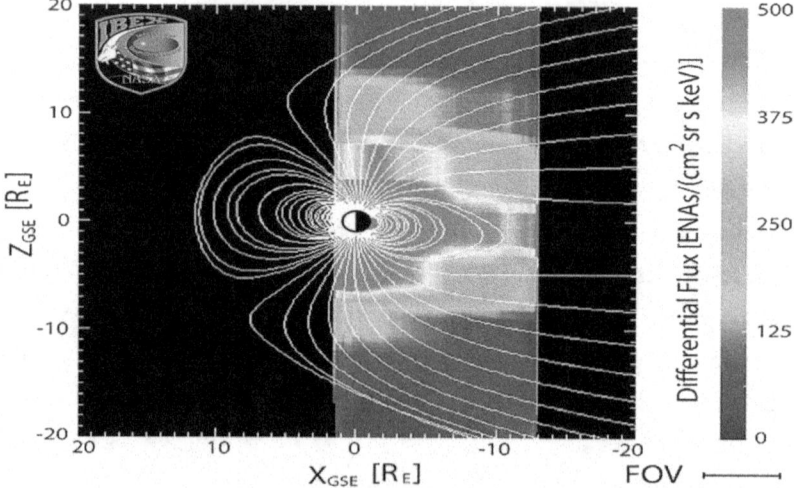

*2a See relevant data provided by Hathaway (MSFC):

Fig.8: shows sunspots form in two bands, one in each hemisphere, that start at about 25° from the equator at the start of a cycle and migrate toward the equator as the cycle progresses...

Fig.13: shows a full-disk magnetogram from NSO/KP used in constructing magnetic synoptic maps over the last two sunspot cycles. Yellows represent magnetic field directed outward. Blues represent magnetic field directed inward.

Fig.14: shows a Magnetic Butterfly Diagram constructed from the longitudinally averaged radial magnetic field obtained from instruments on Kitt Peak and SOHO.

This illustrates Hale's Polarity Laws, Joy's Law, polar field reversals, and the transport of higher latitude magnetic field elements toward the poles.

Fig.20: shows Cosmic Ray flux from the Climax Neutron Monitor and rescaled Sunspot Number. Cosmic ray variations are anti-correlated with solar activity but with differences depending upon the Sun's global magnetic field polarity (A+ indicates periods with positive polarity north pole while A– indicates periods with negative polarity). The positively charged cosmic rays drift in from the heliospheric polar regions when the Sun's north polar field is directed outward (positive). When the Sun's north polar field is directed inward (negative) the positively charged cosmic rays drift inward along the heliospheric current sheet where they are scattered by corrugations in the current sheet and by magnetic clouds from CME's. The negatively charged cosmic rays (electrons) drift inward from directions (polar or equatorial) opposite to the positively charged cosmic rays that are detected by neutron monitors.

Fig.30: shows the absolute asymmetry (North – South) of several key indicators. It is clear from this figure that hemispheric asymmetry is real (it consistently appears in all four indicators) and is often persistent – lasting for many years at a time.

*3 Per data provided by Prof. Subash Kak, Oklahoma State U, it links the outer cosmology (galaxy) with the inner cosmology of living-cell...

*4 Data provided by Prof. Freedman, U. of California (Prof. Young of CMU) shows existence of a 'mysterious dark-object, at the galactic center, of mass 5.2×10^{36} kg or 2.6×10^{6} times the mass of Sun with radius of no more than about 10^{11}m, comparable to AU' (AU $\sim 1.496 \times 10^{11}$ m, meters). In Bose-Einstein condensation 'particles' that naturally repel each other, cooling to a very low temperature, K<2.75 (or in cold, at Neutron) cause them to fall together (or "condense") into a 'new form' and move slower under increased cold-Magnetic force. Heating this 'new form' of matter to a high-temperature, K>5777 (or in hot, at Sun) causes the 'particles' fall-apart and move faster under decreased cold-Magnetic force as some of it is converted into hot-Electric force. That makes hot-matter migrate [see Cals: orbital-speed of hot-matter at Sun ~436,750 m/s] towards cold, and cold-matter towards hot…[see Cals: orbital-speed of cold-matter of Neutron at 39.15 AU from Sun ~97,700 m/s], a steady-flow of dark matter between hot-cold regions. This is similar to circulation of air near a large body of water on a hot day. Here, the hot-Sun, one half-cell, is connected by a conductive electrolyte, dark matter, to the cold-Neutron, the other half-cell, with objects (Planets etc…) in between rotating, orbiting and spinning, and creating electro-magnetic force (emf) in space, like in a battery where the negative and positive electrodes convert chemical energy into electrical energy with anions (negative-ions) and cations (positive-ions) migrating.

*5 Describes force constant of Sun on the Planet, and per 2002 CODATA the Gp is believed to be equal to 6.6742x10^-11 (although the values obtained in experiments varied from 6.6709 to 6.6756x10^-11…)

*6 Per NASA's JPL: "Aside from its galactic orbital velocity (250-300+ km/s), the Sun and its planetary system wander through the local stellar neighborhood at roughly 100,000 km/hr, entering and leaving various tenuous local clouds of gas on a time scale of roughly once every few thousand to millions of years." NASA's wander-speed at 27,700 m/s (~6 times more…) 'once every few thousand to millions of years' …must be erroneous.

*7 Neutron's force, Fis, acting on our-Sun must be the cause of Sun pole flips once every 10-11 years in its sunspot cycle. During the long-term solar activity of grand Maxima-Minima of 942 year-cycle, the force, Fis, increases as the distance, Rs, between our Sun-Neutron decreases (as in the present ½ cycle of 471 years) until both Suns come on to the same side of the Neutron (at 130AU apart). The force, Fis, decreases as the distance, Rs, increases (in the next ½ cycle of 471 years) until our-Sun comes on to the opposite side of Neutron with respect to the companion. Sun inhales cold-matter at the poles and on its surface, and exhales hot-matter at the sunspots and on its surface- as all living do…

In June 2011, Drs. Hill & Penn (of National Solar Observatory), and Dr. Altrock (of USAF Research Labs.), have reported: (i) The specific solar-wind beneath the surface of Sun failed to appear during the present solar-cycle, so next sunspot cycle which is due in 2013 could be delayed to 2021 or may not happen at all. (ii) The magnetic field strength associated with sunspots has been reducing of late. (iii) The process 'rush to the poles' appears to be slowing down with a very weak solar Max. in 2013 (see AAS-Solar Physics Div. meeting in Las Cruces, NM, US). Such conclusions appear to be erroneous based on the 'sunspot' activity witnessed in Feb.-March 2012…

*8 The red-shift (or parallax) phenomena appears to be local in an Atom, and the present-method to calculate the distance between Earth and another Atom or Galaxy using the red-shift in that other Atom appears incorrect. However, we propose an approximate correction-factor to reduce the red-shift distances accounting for the Neutrons/dark matter in space.

*9 The binding and force-transfer between Neutron-Sun-Earth is similar to the force-transfer between the string and an orbiting ball at one-end of the string. Per Prof. Nave, Georgia State U, the string must provide necessary centripetal force to move the ball; if the string breaks the ball will go in a straight line, not in a circular path, as it required string tension. Likewise, there must be a string like action between Neutron-Sun-Earth through dark matter to help transfer forces and keep Earth in its path. We can see the action of dark matter watching high-low tides at Sea, although we cannot see the string.

*10 Per NASA's recent news (Oct. 2009), Fermi Gamma-ray Space Telescope might be able to find dark matter in our galaxy.., as theoreticians have demonstrated that small clumps+ of dark matter in our galaxy… may be more visible than previously thought. Dark matter particles orbiting within self-bound clumps tend to move slower than those bound only to the larger halo that wraps around the entire galaxy… If the theory is on target, the Fermi telescope could detect hundreds+ of small dark matter halos+ in the Milky Way alone… Researchers are still uncertain exactly what dark matter is, but a predominant theory suggests it is made up of super-symmetric particles that act as their own anti-particles… Rather than a handful of small dark matter clusters, Fermi could see hundreds; but it's been a year since data collection began and Fermi hasn't seen any dark matter yet...

*11 Per NASA: Electromagnetic energy decreases as if it were dispersed over the area on an expanding sphere, expressed as 4piD^2 where radius D is the distance the energy has traveled. The amount of energy received at a point on that sphere diminishes as 1/D^2. This relationship is known as the <u>inverse-square law</u> of (electro-magnetic) propagation. It accounts for loss of signal strength over space, called <u>space loss</u>. The inverse-square law is significant to the exploration of the Universe, because it means that the concentration of electro-magnetic radiation decreases very rapidly with increasing distance from the emitter.

*12 Sun emits high-radiation, T ~5777K (at corona, heliopause & pole near Neutron: 1x10^6K). Per Prof.Bennett of JHU, NASA's COBE-WMAP, observations near Earth) found radiation, T~2.7K, at infinite locations in space...

*13 We receive great signals, slowly rising and falling electro-magnetic waves of more than 30x10^9m (or 18x10^6 miles or equivalent to ~4x10^9 billion light years) long streaming from the depths of space. Such faint cosmic flutters are one extreme of the groupings known as radio waves that extend in wavelengths down to 0.3m.

*14 Galaxies can accelerate away from one another if they are an assembly of Cells <u>similar</u> to archaea-bacteria; an asexual, that can divide or split (fission) and undergo sporic meiosis leading to the formation of haploid spores, and able to grow or fuse (fusion) such spores into multi-cellular individuals or dismember and form into a new cell. In biology it says, the living-cells, eukaryotes with nucleus (Dna-Rna), have evolved from prokaryotes with <u>no</u> nucleus. Archaea-bacteria is prokaryote with no nucleus, but similar to eukaryotes; the processes occur across cell membrane where ribosomes (with Dna-Rna) produce proteins, and protein collide to create energy (like GRBs in space). Cell-biology teaches that all living-cells contain cell membrane, ribosomes, Dna-Rna, proteins or enzymes, which make up about 30% of the cell; the balance 70% is cytosol, the conductor of electricity, in which all cell-organs reside (or float).

*15 Per NASA's JPL "The spacecraft Voyager 2 would have fallen back toward the Sun with perihelion at 1 AU, and aphelion at Jupiter's distance of about 5 AU if Jupiter had not been there at the time of spacecraft's arrival, and would have remained in elliptical orbit as long as no other forces acted upon it. …Voyager 2 when it more than doubled its speed with gravity assists in the outer solar system you would feel only a continuous sense of falling. No acceleration."

The spacecraft Voyager 2, if it had the ability to create force or power, like the space station or the satellites orbiting Earth, it would have become a satellite to the nearest-Planet. If it had no ability to create the force it would have crashed into a Planet like an asteroid, but not fallen back toward the Sun or orbit Sun like a Planet…The 'sense of falling' must be due to Neutron's force of gravity on Voyager 2.

*16 Earth orbit Cals: Sun-Neutron forces

From NASA data in Table 1, Earth orbit velocity, V_p=29786m/s, R_{pa}=1.521x10^11m, R_{pp}=1.471x10^11m, Earth mass, P_m=5.974x10^24kg; Sun mass, S_m=1.989x10^30kg

So, from Eq. 1, G_p=[(29786)^2x (1.496x10^11)]/(1.989x10^30)=6.673x10^-11

Then, from Eq. 2,

F_{pp}=6.673x10^-11[(5.974x10^24)(1.989x10^30)/(1.471x10^11)^2]=366.43401x10^20

F_{pa}=6.673x10^-11[(5.974x10^24)(1.989x10^30)/(1.521x10^11)^2]=342.73835x10^20

$(V_{pp})^2$=[(366.43401x10^20)(1.471x10^11)]/(5.974x10^24)=9.0228396x10^8

$(V_{pa})^2$=[(342.73835x10^20)(1.521x10^11)]/(5.974x10^24)=8.7262308x10^8

See Figure 1b-1c:

For Earth: knowing R_{pa}, R_{pp}, P_i=0.00005d; R_s=161x10^11m, Neutron mass, I_m=4x10^30kg; Find: R_{spa}, R_{spp}, A, P

A=0.54127d, Sin A=0.0094468; P=0.52348d, Sin P=0.0091363;

R_{spa}=[(161x10^11)-[(1.521x10^11)(Sin 0.5x10^-4]~161x10^11m;

R_{spp}=[(161x10^11)+[(1.471x10^11)(Sin 0.5x10^-4]~161x10^11m;

Use the above values and Eqs.2P & 2A to determine force constants of Planet, B_{sp} & B_{ip} as follows:

Where, Bsp is due to Sun-force, and Bip is due to Neutron-force on the Planet-

$9.0228396 \times 10^8 = Bsp[(1.989 \times 10^{30})/(1.471 \times 10^{11})] + Bip[(4 \times 10^{30})(0.0091363)/(161 \times 10^{11})]$

$\qquad = Bsp(1.3521414 \times 10^{19}) + Bip(0.000227 \times 10^{19}) \ldots \ldots \ldots (1)$ at perihelion

$8.7262308 \times 10^8 = Bsp[(1.989 \times 10^{30})/(1.521 \times 10^{11})] - Bip[(4 \times 10^{30})(0.0094468)/(161 \times 10^{11})]$

$\qquad = Bsp(1.3076923 \times 10^{19}) - Bip(0.0002347 \times 10^{19}) \ldots \ldots \ldots (2)$ at aphelion

Solve (1)+(2) and assume Bsp+Bip~ Bsp (like Newton's Gp)

Then we get, $17.74907 \times 10^8 = Bsp(2.659826 \times 10^{19})$

$$Bsp = 6.673 \times 10^{-11}$$

From (1), $(Vip)^2 = (17.74907 \times 10^8)(0.000227 \times 10^{19})/(2.659826 \times 10^{19}) = 1.514108 \times 10^5$

From Eq.2a, $= (FippX)(Rspp)/Pm;$

Or $FippX = (1.514108 \times 10^5)(Pm)/(Rspp) = Bip(Im. Pm)(Rspp)^2$ see Eq.2d

So, $Bip = (1.514108 \times 10^5)(161.00672 \times 10^{11})/(4 \times 10^{30}) = 0.06095 \times 10^{-11}$

From (2), determine at aphelion $Bsp = 6.673 \times 10^{-11}$ & $Bip = -0.06362$: Follow the same procedure for all Planets

Orbit velocity from Eq.2D

$Vp = \frac{1}{2}\{[(Bspp)(Sm/Rpp)]^{\frac{1}{2}} + [(Bspa)(Sm/Rpa)]^{\frac{1}{2}} + [(Bipp)(Im/Rspp)]^{\frac{1}{2}} - [(Bipa)(Im/Rspa)]^{\frac{1}{2}}\}$

$$Vp \sim \frac{1}{2}\{[(6.673 \times 10^{-11})(1.989 \times 10^{30}/1.471 \times 10^{11})]^{\frac{1}{2}}$$
$$+ [(6.673 \times 10^{-11})(1.989 \times 10^{30}/1.521 \times 10^{11})]^{\frac{1}{2}}$$
$$+ [(0.06095 \times 10^{-11})(4 \times 10^{30}/161 \times 10^{11})]^{\frac{1}{2}}$$
$$- [(0.06362 \times 10^{-11})(4 \times 10^{30}/161 \times 10^{11})]^{\frac{1}{2}}\}$$
$$= 29{,}786 \text{ m/s}$$

Conclusion: As Rpp and Rpa increase, Bip increases more rapidly than Bsp: See Table 2 & Fig.2.

Take (2) as an example, and calculate the forces acting on Earth at aphelion (XY plane):

$(Vsp)^2 = [(8.7262308 \times 10^8)(1.3076923 \times 10^{19})]/[(1.3076923 \times 10^{19}) - (0.0002347 \times 10^{19})]$

$\qquad = (8.7277972 \times 10^8) = (Fspa)(Rpa)/Pm$ See Eq.2a

Or Fspa=(8.7277972x10^8)(5.974x10^24)/(1.521x10^11) ~342.8x10^20kg

Sun Z-force (FsZ) at aphelion & perihelion of Planet that causes wobble-spin of ZZ,

FsZpa=½(FspaX+FsppX)(Sin Pi), Pi=0.00005, use FspaX ~FsppX See Eq.2D
  =Fspa (Sin Pi)~ (342.8x10^20)(0.000000873)=2.9915x10^16 kg

Note: Use FsZpa~FsZpp that acts on YY along Z; at equinoxes
FsZpa=FsZpp=0 or no Z-force since Pi=0. So, 2.9915x10^16 (at aphelion or
perihelion) becomes 0 (at equinoxes). The drag-force created by Sun Z-force
causes Earth ZZ-axis to tilt (from YY, wobble) and spin anti-clockwise. Use
the net-effect of FsZ for 64 days out of 365 days in each orbit of the Earth,
with high or low drag force constant, C1 or 0.25C1

(Vip)^2=[(8.7262308x10^8)(0.0002347x10^19)]/[(1.3076923x10^19)-(0.0002347x10^19)]

  =(0.0015664x10^8)=(FiXpa)(Rspa)/Pm   See Eq.2a

  FiXpa=(0.0015664x10^8)(5.974x10^24)/(161x10^11)] ~5.812x10^16 kg

  Fipa~ (5.812x10^16)/(Sin A) =615.26x10^16kg.   See Eq.2d

  FiZpa= (Fipa)(Cos A)=615.26x10^16 kg

Note: Use FiXpa~FiXpp, that acts in X-direction on Earth's diameter (YY)
and creates drag force to spin ZZ clockwise.

Indra Z-force (FiZpa) creates Y-force at equinox of Earth and causes tilt of
ZZ (up or down) to wobble
FizY= ½(FiZpa+FiZpp)(Sin ½(A+P)), use FiZpa ~FiZpp   See Eq.2D
  = (615.26x10^16)(0.0092916)= 5.7162x10^16 kg

Note: FizY becomes 0 at aphelion and perihelion; so, the tilt of ZZ varies from
0 to Max. at equinoxes. Using Sun's FsZ and *Indra* FizY, one can calculate the
up and/or down motions of Earth ZZ-axis (in YY) as part of the wobble, while
Indra X-force (or its drag force) causes Planet ZZ spin clockwise.

*16.1 Sun Y-force that rotates Earth at Ai=23.5E,

FsY= ½[FspaY+FsppY]C* See Eq.2D

*C=Constant of interaction between Sun-Earth magnetic field-intensity, assume—
  C=1 to 0.9, at Ai=0N-180S, since ZZ-axis stays closer to <u>high</u> magnetic flow; and
  C=0 at Ai=0N, Ai=90E, Ai=180S & Ai=270W, since the interaction will be
  none at these locations.

Force couple that rotates Earth~ (FsY x Earth radius)
~(0.9x342.8x10^20kg)(*6378140m)=1.96776x10^29 kg m
Speed of rotation, (Ver)^2=(1.96776x10^29/Mer=(465)^2m/s
Mer=(1.96776x10^29)/(465x465)= 0.91x10^24 kg
*Approximate with rotating mass at surface
Mer=Earth mass (mantle+crust) rotates ~0.91x10^24kg:
Convert rotating mass into equal-thick, t=(6378140-r)
Find r from [(*35.35x342/2.2)(4pi/3)][(6378140)^3– (r^3)]=0.91x10^24 kg;
We get, r=6036200 m *density kg/m^3
That gives a thickness of (mantle+crust) the rotating mass,
Met= (6378140- 6036200) ~342 km

<u>From Eq.2D</u>:
Indra X-force spins ZZ clockwise (Ai=180S-225W or 0N-45E),
                    FiX1=[½(FipaX+FippX)]C1*
*C1=(C)(Sin Ai)(Sin Ai x Cos Ai)^2, Ai=0-45;
  C1=<u>high</u> drag-force constant deduced from estimated IERS Zs23.5
~0.2arcsec/Yr., see below.

Indra X-force spins ZZ clockwise (Ai=225W-270W or 45E-90E),
                    FiX1=[½(FipaX+FippX)]C1*
*C1=(C)(Sin 45)(Sin 45 x Cos 45)^2, Ai=45;
  C1=constant of <u>high</u> drag-force at Ai=45 remains the same until ZZ gets to
90E, because during this time ZZ will be in front of magnetic-N at <u>high</u> flow-
intensity (between 225W~270W).

Indra X-force spins ZZ clockwise (Ai=90E-135E),

$$FiX2=[½(FipaX+FippX)]C2*$$
*C2=(0.25C)(Sin Ai)(Sin Ai x Cos Ai)^2, A=0-45; C2=<u>low</u> drag force constant

Indra X-force spins ZZ clockwise (Ai=135E-167E),
$$FiX2=[½(FipaX+FippX)]C2*$$
*C2=(0.25C)(Sin Ai)(Sin Ai x Cos Ai)^2, A=45-77; C2=<u>low</u> drag force constant.

Indra X-force spins ZZ clockwise (Ai=167E-180S),
$$FiX2=[½(FipaX+FippX)]C2*$$
*C2=(0.25C)(Sin 77)(Sin 77 x Cos 77)^2, A=77;

  C2=<u>low</u> drag force constant- with ZZ at <u>high</u> flow-intensity of magnetic-N beyond Ai=77 (or 167E ~180S) the constant C2 remains the same as at Ai=77.

<u>Note</u>:FiX1-FiX2 act in XY-plane on Planet-diameter (YY) causes clockwise spin of ZZ

<u>Earth clockwise-spin, Ai=0N-90E</u>: (past, present & near-future) Per IERS Zs23.5 ~0.2 arcsec/Yr.(Approx. or close to 0.195?)
So, (Ves)^2=[(Earth Radius x 0.2)/(365x24x3600)]^2=0.001636
    (Ves)^2= F.R/Mer = 0.001636 R=Radius
Force that causes clockwise spin, F=[(0.001636)(0.91x10^24)]/(6378140)
                    =2.3344x10^14 kg
At Ai=0N-45E,
F=[(FiX1)C1]−{(64/365)x½[(FsZpax0.25C1)+(FsZppxC1)]}, use FsZpa ~FsZpp

<u>Note</u>: C1=<u>high</u> drag-force constant going <u>against</u> magnetic-flow, and 0.25C1 for <u>low</u> drag-force constant going <u>with</u> magnetic-flow, an estimated net-effect of Sun's FsZ at aphelion-perihelion in each orbit= 64 days out of 365 days.

F=[(FiX1)C1]−{(64/365)x 5/4 (FsZ)C1}
  =5.812x10^16C1−0.2191781x2.9915x10^16C1=2.3344x10^14
  =5.812x10^16C1−0.6556x10^16C1=2.3344x10^14

$$C1=(2.3344x10^{\wedge}14)/(5.15633x10^{\wedge}16)=0.004527$$

Or, if we put the C1 in terms of variable Ai, at Ai=23.5,

$$C1=(9x10^{\wedge}-2)(\text{Sin Ai})(\text{Sin Ai} \times \text{Cos Ai})^{\wedge}2$$
$$=(0.09)(0.05332)= 0.0045 \sim 0.004527$$

Therefore, at Ai=45E (or 225W),

F45=[(0.09)(Sin 45)(Sin 45xCos 45)^2][(5.812x10^16)-(0.6556x10^16)]
    =8.204x10^14 kg

(Ves)^2=F.R/Mer=(8.204x10^14)(6378140)/(0.91x10^24)= 0.005754

Ves=0.075855 m/s

Or, spin speed, Vs=[(0.075855x365x24x3600)/(6378140)]=0.375 arcsec/Yr.

Conclusion: High drag-force constant varies at-

C1=(0.09)(Sin Ai)(Sin Ai x Cos Ai)^2, between Ai=0-45 (0N-45E); and

C1=(0.09)(Sin 45)(Sin 45 x Cos 45)^2, between Ai=45-90 (45E-90E).

Note: At Ai=45E-90E, Earth ZZ will be at high flow intensity being pushed forward by magnetic-N; therefore, C1 will remain constant (no change) beyond Ai=45

So, this meant Earth period of spin-

From Ai=0N-45E, *T3*=(45x 3600)/(0+0.375)/2=864,000 years

From Ai=45E-90E, *T4*=(45 x 3600)/(0.375) =   432,000 years

Earth clockwise-spin, Ai=90E-180S: (past or future)

Low drag-force constant varies as above-

C2=(0.25C)(Sin Ai)(Sin Ai x Cos Ai)^2, between Ai=0-77 (90E-167E)

Or, C2=(0.0225)(Sin Ai)(Sin Ai x Cos Ai)^2; and

C2=(0.0225)(Sin 77)(Sin 77 x Cos 77)^2, between Ai=77-90 (167E-180S).

Therefore, at Ai=45 (or 135E),

F45=[(0.0225)(Sin 45)(Sin 45xCos 45)^2][(5.812x10^16)-(0.6556x10^16)]
    =2.051x10^14 kg

$(Ves)^2 = F.R/Mer = (2.051 \times 10^{14})(6378140)/(0.91 \times 10^{24}) = 0.0014375$

$Ves = 0.03791 m/s$

Or, spin speed, $Vs = [(0.03791 \times 365 \times 24 \times 3600)/(6378140)] = 0.1875$ arcsec/Yr.

And, at Ai=77 (or 167E),

$F77 = [(0.0225)(Sin\ 77)(Sin\ 77 \times Cos\ 77)^2][(5.812 \times 10^{16}) - (0.6556 \times 10^{16})]$

$= 0.5431 \times 10^{14}$ kg

$(Ves)^2 = F.R/Mer = (0.5431 \times 10^{14})(6378140)/(0.91 \times 10^{24}) = 0.0003807$

$Ves = 0.01951 m/s$

Or, spin speed, $Vs = [(0.01951 \times 365 \times 24 \times 3600)/(6378140)] = 0.0965$ arcsec/Yr.

So, this <u>meant</u> Earth period of spin-

From Ai=0N-45E, $T1 = (45 \times 3600)/(0 + 0.1875)/2$	=<u>1,728,000</u> years
From Ai=45E-90E, $T2 = (32 \times 3600)/(0.1875 + 0.0965)/2$	= 811,100
$(13 \times 3600)/(0.0965)$	= 484,900
	<u>1,296,000</u> years

Yuga is 4, 320,000 years (1.728+1.296+0.864+0.432)

*16 Note

(1) **From:** Mikki

Sent: Sunday, February 26, 2012 8:37 PM

To: 'Christian Bizouard'

Cc: 'Ulrich Schreiber'; 'prd@ridge.aps.org'; 'prl@ridge.aps.org'

Subject: Earth's polar motion

02/26/2012

Dear Dr.Bizouard:

I want to thank you for prompt responses you sent me during the past 2-weeks (although couple of issues unresolved…).

I am pleased to attach a copy of my final product (in the form of a Table) which I prepared using IERS data on Earth's polar motion (X, Y in arc sec).

In reality, there is a good reason for the Wobble radii Ra to occur in June-July of the year (at aphelion) & Rp in Dec.-Jan. (at perihelion) with Ra, Rp being close to zero at equinoxes... But, per IERS data the Wobble radii Ra, Rp occurs in different month every year and not always in June-July & Dec.-Jan.

You can see that only occurs once every 3 or 4 years apart. That makes no sense to me !

Therefore, as you can see (at the bottom of the Table)- I made the conclusions using the Wobble radii of select years, i.e. during the years when Ra occurred in June-July & Rp in Dec.-Jan. of the year.

Please, also, note that Earth's polar motion (or drift) works out to be ~0.015 arc sec/Yr., and it is ~5 times greater than IERS table (Fig.4) which is ~0.00323 arc sec/Yr.

As usual, please feel free to write to me with your comments or Qs...

This is the last phase, an important part, of my work...

Thanks, Maheswar

(2) **From:** Mikki
Sent: Friday, February 24, 2012 12:28 PM
To: 'Christian Bizouard'
Subject: Earth's polar drift?
02/24/12

The description on VLBI below is from NASA.

It says 'quasars' define *celestial reference frame*,...a true inertial reference frame... that their motions across the sky are <u>undetectable</u>.... This is the same *reference frame used for measuring Earth orientation.*

This "<u>undetectable</u>" does not mean we should accept as true with closed 'eyes'- recall Nobel prize of 2011- the so called quasars too accelerating if you believe those who received Nobel? So, I submit there is a fallacy in the "*reference frame used for measuring Earth orientation*".

Also, we have no VLBI at or close to +90 latitude or at Earth's axis of rotation... (although the VLBI near the equator in S. America may yield reasonably good data)... Therefore, I am not sure how reliable is IERS data?

I hope you or someone at IERS would look into this and other issues I raised and get back with me...

(3) **From:** Mikki
Sent: Monday, November 28, 2011 9:44 AM
To: 'Christian Bizouard'
Cc: 'Ulrich Schreiber'; 'prd@ridge.aps.org'; 'prl@ridge.aps.org'
Subject: IERS data
11/28/2011

Dear Dr.Bizouard:
Thank you for your prompt response.

I agree with you in principle, 'Earth's pole drift' is a reality.
Also, I agree with you "the accuracy of the pole drift depends on the accuracy/ stability of the measurements..."
But, unfortunately, we will not be able to determine that 'pole drift' accurately the way IERS is doing- taking distant "stars" or "galaxies" as reference points on an imagined base-line.
Why? As you know, the stars or the points on your base-line move- keep moving in different directions at different speeds...

So, the imagined points on your base-line do not stay fixed...to allow you to determine Earth's actual drift (if you allow me- that's exactly what happens in any living entity- you or I or the tree; all living atoms and the particles within keep moving at variety of speeds, like Mercury-Pluto etc...
I know we can determine that "drift" or continued 'tilt' by other means- it is not easy to explain that here, so please wait until my work to publish.
Regarding "Earth's continued polar motion can be estimated assuming the imagined Origin continues to move (not fixed)"- please allow me to explain:
Assume for our purpose IERS observed data defines the polar motion of 'a point' on Earth's axis of rotation in terms of X-Y relative to the reference points on the imagined base line (in space). If so, project that 'imagined base line' on to the

Earth as X=0, Y=0. Then, the observed 'point' on the Earth must be moving in circles (or ellipses) with reference to this fixed base line... We know in reality the base line is not fixed, but keeps moving, and we have no way of determining that...Correct?

Therefore, IERS observed data or Dr.Schreiber experiment can only tell us how the 'point' on the Earth moves... (in reference to an imagined base-point at X=0, Y=0, which itself is a moving target- as you call it a 'drift'). I know we cannot design an experiment that can determine this 'drift' accurately.

So, the Q- is there a use for IERS observed data on polar motion of a point on the Earth in terms of X, Y?

Yes, there is- it can help solve the big puzzle if we know more about the reality... As I reported the IERS data gives an average polar motion of the 'point': 0.17 arcsec/year (minimum), and 0.49 arcses/year (maximum)- the mean of these values being at ~0.33 arcsec/year.

That tells me- the 'point' on Earth's axis of rotation is drifting now at ~0.33 arcsec/year. Does that mean the imagined X=0, Y=0 is also drifting at that rate? That I may be able to answer- soon !

Maheswar

——Original Message——
From: christian bizouard [mailto:christian.bizouard@obspm.fr]
Sent: Monday, November 28, 2011 5:29 AM
To: Mikki
Cc: Christian Bizouard; 'Ulrich Schreiber'; prd@ridge.aps.org; prl@ridge.aps.org
Subject: Re: IERS data

Dear Mikki,

Please apologize the delay of my answer to your letter of November 14th. It was covered by a geological strata of email, and I had forgotten it.

"Earth's continued polar motion can be estimated assuming the imagined Origin continues to move (not fixed)". What do you mean by "imagined" Origin? I am confused...The geographic pole? Of course the accuracy of the pole drift depends on the accuracy/stability of the measurements which were employed from the years 1900. But I guess that since the 1970's, this drift is well determined and confirms early astrometric observations.

Best regards,

Christian B.

*16.2 Moon orbit Cals: Earth-Neutron force (Sun has no direct-affect)
From NASA data in Table 1, Moon's orbit velocity, $Vm=1023m/s$,
$Rma=4.055x10^8m$, $Rmp=3.633x10^8m$; Earth mass, $Pm=5.974x10^{24}kg$;
Moon mass $Mm=0.73483x10^{23}kg$

So, from Eq. 1,
$Bm=[(1023)^2$ x $(3.844x10^8)]/(5.974x10^{24})=6.734x10^{-11}$

Then, from Eq. 2,
$Fmp=6.734x10^{-11}[(5.974x10^{24})(0.73483x10^{23})/(3.633x10^8)^2]=2.2397x10^{20}$
$Fma=6.734x10^{-11}[(5.974x10^{24})(0.73483x10^{23})/(4.055x10^8)^2]=1.7978x10^{20}$
$(Vmp)^2=[(2.2397x10^{20})(3.633x10^8)]/(0.73483x10^{23})=1.1073195x10^6$
$(Vma)^2=[(1.7978x10^{20})(4.055x10^8)]/(0.73483x10^{23})=0.9920818x10^6$

See Figures 1b-c:

At Earth's aphelion: knowing Rma, Rmp, Rs, Im, $Pi=0.5x10^{-4}d$, $Mi=5.145d$,

Ai~23.5d: Calculate- Rsma, Rsmp, A', P'

A'=0.001444d, Sin A'=0.0000252; P'=0.001295d, Sin P'=0.0000226

      $Rspa=[(161x10^{11})-[(1.521x10^{11})(Sin 0.5x10^{-4})\sim 161x10^{11}m$

      Rma (Sin Mi)=0.00194x10^11;

      Rmp (Sin Mi)=0.00174x10^11

      Rsma=(Rspa -0.00194x10^11)= 160.998x10^11m

      Rsmp=(Rspa +0.00174x10^11)=161.00174x10^11m

For Moon at perigee: use the above values and Eq.2P to determine force constants of Moon, Bpm & Bim as follows:

Where, Bpm is due to Planet force, and Bim is due to Neutron force on the Moon: When Earth is at its perihelion:

$1.1073195 \times 10^{\wedge}6 = Bpm[(5.974 \times 10^{\wedge}24)/(0.003633 \times 10^{\wedge}11)] + Bim[(4 \times 10^{\wedge}30)(0.0000226)/(161.00174 \times 10^{\wedge}11)]$

$= Bpm(1.644371 \times 10^{\wedge}16) + Bim(0.0005615 \times 10^{\wedge}16)$..............(1) at perigee

$0.9920818 \times 10^{\wedge}6 = Bpm[(5.974 \times 10^{\wedge}24)/(0.004055 \times 10^{\wedge}11)] - Bim[(4 \times 10^{\wedge}30)(0.0000252)/(160.998 \times 10^{\wedge}11)]$

$= Bpm(1.4732429 \times 10^{\wedge}16) - Bim(0.0006261 \times 10^{\wedge}16)$............ ..(2) at apogee

Solve (1)+(2) and assume Bpm+Bim~ Bpm (like Newton's Gm)

Then we get, $2.0994012 \times 10^{\wedge}6 = Bpm(3.11755 \times 10^{\wedge}16)$

$$Bpm = 6.7341 \times 10^{\wedge}-11$$

From (1), $(Vim)^{\wedge}2 = (2.0994012 \times 10^{\wedge}6)(0.0005615 \times 10^{\wedge}16)/$

$$(3.11755 \times 10^{\wedge}16) = 0.0003781 \times 10^{\wedge}6$$

$= (Fimp)(Rsmp)/Mm,$ see Eq.2a

Or $Fimp = (0.0003781 \times 10^{\wedge}6)(Mm)/(Rsmp)$

$= Bim(Im. \ Mm)(Rsmp)^{\wedge}2$ see Eq.2

So, $Bim = (0.0003781 \times 10^{\wedge}6)(161.00174 \times 10^{\wedge}11)/(4 \times 10^{\wedge}30) = 0.0001522 \times 10^{\wedge}-11$

From (2), determine at apogee $Bpm = 6.7341 \times 10^{\wedge}-11$ & $Bim = 0.0001697 \times 10^{\wedge}-11$,

And follow the same procedure for all Moons.

<u>Conclusion</u>: As Rmp and Rma increases, Bim increases more rapidly than Bpm and that results in retrograde orbit.

Take (1) as an example, and calculate <u>forces acting on the Moon at perigee</u> (in XY plane):

$(Vpmp)^{\wedge}2 = [(1.1073195 \times 10^{\wedge}6)(1.644371 \times 10^{\wedge}16)]/[(1.644371 \times 10^{\wedge}16) + (0.0005615 \times 10^{\wedge}16)]$

$= (1.1069415 \times 10^{\wedge}6) = (Fpmp)(Rmp)/Mm$ See Eq.2a

Or $Fpmp = (1.1069415 \times 10^{\wedge}6)(0.73483 \times 10^{\wedge}23)/(3.633 \times 10^{\wedge}8) = 2.239 \times 10^{\wedge}20kg$

  $FpXmp = Fpmp = 2.239 \times 10^{\wedge}20kg$

$(Vimp)^{\wedge}2 = [(1.1073195 \times 10^{\wedge}6)(0.0005615 \times 10^{\wedge}16)]/[1.644371 \times 10^{\wedge}16) + (0.0005615 \times 10^{\wedge}16)]$

$= (0.000378 \times 10^{\wedge}6) = (FiXmp)(Rsmp)/Mm$ See Eq.2a

Or $FiXmp = (0.000378 \times 10^{\wedge}6)(0.73483 \times 10^{\wedge}23)/(161.00174 \times 10^{\wedge}11) = 1.7253 \times 10^{\wedge}12kg$

  $Fimp = FiXmp/Sin \ P' = (1.7253 \times 10^{\wedge}12)/(0.0000226) = 7.634 \times 10^{\wedge}16kg$ See Eq.2d

FiZmp = (Fimp)(Cos P')= 7.634x10^16 kg   See Eq.2d

FiXmp = Fimp (Sin P') = 1.7253x10^12kg

Force that makes Moon orbit (rotate),

FpYmp=Fpmp=2.239x10^20kg

Force that makes Moon orbit-inclination,

FiZmp=Fimp (Cos P')=7.634x10^16kg

Conclusion: Ei is the angle at which Planet radial-force acts along Moon's orbit- neglect it being small. Take all forces act on the surface of the object.
(i) Planet X-force (FpmX) causes Moon to orbit, and Neutron Z- force (FimZ) causes orbit-inclination, Mi. As the distance, Rmp, increases to Rma Planet X-force (FpmX) reduces; an increase in Rma increases Neutron Z-force (FimZ)- that increases Mi and causes retrograde orbit to the outer Moon...
(ii) Planet X-force (FpmX) acts on the surface and rotates the Moon.
(iii) Neutron X-force (FimX) acts on the surface and spins Moon ZZ-axis...

*16.2 Note
(1) From: Mikki
Sent: Thursday, January 26, 2012 11:17 AM
To: 'zuber@mit.edu'
Cc: 'P. Kenneth Seidelmann'
Subject: Moon?

01/26/2012
Hi, Prof.Zuber and Prof.Seidelmann:

I wish Prof.Zuber would refer to my earlier e-mail of 01/02/2012 (below) and answer a simple-Q I have:
Please see the Fig. below (copied from Wikipedia) and advise if your experiment can confirm my finding that at the present time Moon's S.Magnetic pole points toward the Earth....?

Of course, I will be delighted to hear from Prof.Seidelmann with an answer...
Mikkilineni

An illustration demonstrating simple synchronous rotation. As the moon takes exactly one orbit to rotate one about its axis, the inhabitants of the planet will never be able to see the green side of the moon.

Synchronous rotation

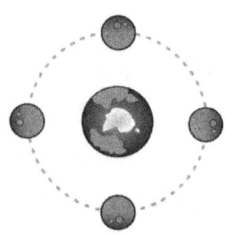

(2) **From:** Mikki
Sent: Monday, January 02, 2012 8:36 AM
To: 'zuber@mit.edu'
Subject: Moon?
01/02/2012

Dear Prof.Zuber:
 Article in NYT said "The side of the Moon that faces Earth is flat and mostly smooth. The other side is mountainous. So, to me, said Dr. Zuber, a professor of geophysics at the Massachusetts Institute of Technology, "if you've taken a hundred spacecrafts' worth of measurements and you still don't know the answer to something fundamental like that, then it's likely the answer isn't on the surface, it's somewhere else. We're making a bet
here, by really learning the internal structure very well, that we'll be able to answer those questions."

This $496 million NASA mission, called the <u>Gravity Recovery and Interior</u>

Laboratory, or Grail, will conduct a single measurement: a map of the Moon's gravitational field. But it will do it with such precision that scientists will get a clear picture of the interior...."

May I submit it would be nice if you can learn 'a clear picture of the interior', but I doubt if it is possible by going around outside... Of course, we have no choice... On the issue of why 'the Moon that faces Earth is flat and the other side is mountainous' I think, I have an answer- *it is the external force that acts on the Moon that makes the orbit-inclination...* To understand this one has to review my calculations...

Thank you, Maheswar

References: None other than cited in the Text and Endnotes

Appendix

A1
From: Mikki
Sent: Friday, November 23, 2012 3:36 PM
To: lars.heikensten@nobel.se; charles.bolden@nasa.gov
Cc: info@barackobama.com; Clinton Foundation (enews@
clintonfoundation.org)....
Subject: Where is Holy Spirit or Creator?
11/23/12

Dear Nobel F & NASA staff-members: [copy of this message sent
to Whitehouse with a request to ask President's science advisor to
contact me...]

On 08/17/12, I wrote to you with a copy of 62-page article on my
work (includes 5 Figures in color)- pleased to attach a copy below.

And, on 09/13/12, I again wrote to you with a copy of Preface and
said:

"*Veda* is pure Science, not a philosophy or mythology. The Books
written and the various thought-schools missed the point. Only
Shankara's *Advaita* (non-duality) is the Reality (which Buddha confirms
indirectly). Please read the Preface to a Book, coming soon..."

I am writing to you today- the Book is ready and expected to be
published before January 2013. Please see the attachment- cover
page, preface & *Veda* symbols that describes the Atom in which we

live of *Brahman* or you may call Father…And, the creator *Isvar* or holy Spirit.

Please note that during the past 400 years, only Kepler and Newton came close to Reality (although they missed it). The rest, including Einstein, are wrong on cosmology or astronomy.

As an example see the latest (which I incorporated into my article):

"J. X. Mitrovica et.al. "Mechanisms for oscillatory true polar wander" (see Nature 491, 244-248, Nov 08, 2012) and "How Earth's wandering poles return home" (see Physicsworld.com, Nov. 12, 2012).

The experts report that the polar wander is the relative movement between Earth surface and its axis. Their computer model predicts the true polar wander that Earth surface tips over and then returns to its original position. "If I sit at the pole, I see the pole shift up to 50° and then turn around, a process of oscillatory true polar wander…I am really surprised that it took about 10X10^6 years to pull and push the poles." When rock cools in a magnetic field it records the magnetic properties of the field; and millions of years later the magnetism in the rocks are decoded in the lab to measure changes in the orientation of the Earth magnetic field stored in ancient rocks. That shows the effects of the oscillatory polar wander.

Cause: See a message of Nov. 20, 2012 to Prof. Mitrovica-

I read the Abstract in Nature on your work- Polar Wander. I also read your comments in PhysicsWorld. I agree it is an important topic to study. I am sure, you are aware of IERS work or measurements on Earth polar wander (including wobble of the axis).

I am an old retired engineer; and I spent lot of time on this topic. I have a complete solution to the topic, or should I say 100% solid proof that there is NO polar wander.

Yes, pole reversal is the Reality every 4.32x10^6 years- precisely. I will be happy to send you my calculations if you agree to give a careful review and willing to publish it in Nature.

I am sure we can make that deal in the interest of Man and Science. With your permission, I can also say this- "If you sit in Boston or at the pole for 10 million years, Boston or the pole would not wander 50-70 degrees and then return close to that location".

And, your concept that the "bulge" and "Plate-stresses" counter the polar wander and make the pole return close to its original position is Flawed. That's not the Reality. Please keep in touch- so, we can learn the Reality."

And, this message will be included as a part of the Book in the Appendix for the benefit of the future generations...

Thanks,

Maheswar

A2

From: Mikki
Sent: Friday, August 17, 2012 6:41 AM
To: 'lars.heikensten@nobel.se'; 'charles.bolden@nasa.gov'
Subject: god?

08/17/12

Dear Mr. Heikensten and Mr. Bolden:

Where is 'god'? We live within 'god' or god's Atom.

You and I know the Tree, you, I, the Earth or Sun, all- we can sense is made of atoms; nothing, but atoms. So, we must try to know the atom…, and one way to know that is to know the Atom of 'god'.

Here is the work I have been doing during the past >10 years. See the attachment (below). Before, you or your experts dive into the Method, please read the Background and Endnotes, and view the Figures to get a picture of- where we went wrong or how to come back into the right path…

Please STOP giving Nobel Prizes in Physics (or related discovery) or spend Billions on useless Research, until you appoint a Panel of Experts and complete review of my work…, if interested.

Why did I pick you to send this message? Because your Nobel Prize is a contributing factor for us to get into this deep black hole we are in.

Please do not dynamite *Veda*, knowledge, our-godly ancients have given us over a million years ago to follow… and don't confuse with the color of skin.

I am close to finishing a Book and soon it will be ready for publication. This message to you will be a part of that Book.

Please feel free to write to me if you have any Qs…

Thanks, Maheswar

A.3
From: Mikki
Sent: Friday, March 15, 2013 3:19 PM
To: 'incandel@hep.ucsb.edu'
Cc: Achintya Rao (achintya.rao@cern.ch); lars.heikensten@
nobel.se; charles.bolden@nasa.gov; info@barackobama.com;
'SpeakerBoehner@mail.house.gov' (SpeakerBoehner@mail.house.gov) …
Subject: god like particle?
03/15/13

Dear Professor

Here we go, again. I remember you were a non-believer or a skeptic
when you took over this present position.

Now, you are made the spokesman for seeing the 'god particle'.
Congratulations.

I saw your presentation. Assuming the test or proton crash data
is correct, I do not know how you could see the so called 'bump'
knowing the protons in the jet flew at varied speeds in variety of
angles at the instant they crashed into one another?

The 'god particle' is the "neutron" which Chadwick discovered
(without seeing or indirectly) and estimated it's mass at ~proton
mass x 1.0067.

Now, you and other 3,000 scientists see a 'god particle' mass at
~proton mass x 125.

And, I estimate the Neutron Mass of the Atom in which we live at
~Sun's Mass x 2x1.006 (which is very close to Chadwick's estimate).

So, we should wonder What is going on or What is the Reality?

See your statement in todays' NYT:

"The preliminary results with the full 2012 data set are magnificent," Joe Incandela, a professor at the University of California, Santa Barbara, and leader of one of the discovery teams, said in a statement released by CERN. "To me it is clear that we are dealing with a Higgs boson, though we still have a long way to go to know what kind of Higgs boson it is."

I asked you earlier a simple Q which you did not answer or maybe you have no answer. I must conclude 'there must be something wrong either with your experiment or use of the vast data'. You can't see a neutron ~125 times massive than a proton when you crash 2-protons, unless the crash is "fusing" 125 neutrons together as a result of the experiment, the jet and the speed of the protons?

If you tell me in simple 'words'- how you make the protons crash, at what speed and the size of the jet etc… (may be) I can help. ?

You know all protons in a jet would not travel at a single central speed- So, you can't control or know the speed or the direction in which any two protons travel or hit one another or into a group? Therefore, the experiment itself must be flawed.

Thanks,

Maheswar

A4
From: Mikki
Sent: Friday, March 22, 2013 12:12 PM
To: lars.heikensten@nobel.se
Sent: Friday, March 22, 2013 8:58 AM
To: 'Richard.Battye@manchester.ac.uk'
Cc: gpe@ast.cam.ac.uk; 'incandel@hep.ucsb.edu'; charles.bolden@
nasa.gov; info@barackobama.com
Subject: Mystery?
03/22/13

Hi, Professor Battye: [with attachment]
This is a smart comment I heard in many years! Thanks.

"This is the first time that man can look with such clearness at the
origin of the Universe, in which we see the impact of forms of matter
and energy still unknown today," cosmologist Carlo Baccigalupi
commented in a statement.

Richard Battye of the University of Manchester's astrophysics
center added: "There are... a number of tantalizing hints that there
is something missing in our understanding. It will be fun trying to
figure out what is going on."

And, the correct answer to this ""As the Universe expanded, these light
waves stretched out into shorter microwave wavelengths, reaching
a temperature today of just 2.7 degrees above absolute zero—minus
273.15 C (minus 459.67 F)" is Not the big-Bang: it is the left over
radiation of the Sun and Stars at their Neutrons (in *Sanskrit* it is known
as *Indra*s: Ancients knew this before the last Ice-age).

So, let us forget the Fiction and get back to the Reality. Please
contact me if you want to know the Reality.

Maheswar

A5

[Attachment]

From: Mikki

Sent: Thursday, March 21, 2013 8:38 AM

To: 'gpe@ast.cam.ac.uk'

Cc: 'incandel@hep.ucsb.edu'

Subject: Mystery?

03/21/13

Hi, Professor:

I read a news item where you said among other things:

[Planck can produce cosmological maps with three times the
resolution of WMAP, and at least 10 times the temperature
sensitivity. As a result, the estimates of the universe's age
and composition have undergone some additional fine tuning.
Efstathiou said Planck's estimate for the age of the universe is
13.81 billion years, plus or minus 50 million years. The Planck
team's breakdown of the universe's constituents is 4.9 percent
ordinary matter, 26.8 percent dark matter and 68.3 percent dark
energy, he said.

"There's less stuff that we don't understand, by a tiny amount,"
Efstathiou said.

He said the Planck data also pointed to some "strange features" in
the universe's temperature distribution, including an unexplained dip
at one point of the power spectrum, and an unusual distribution of
large-scale fluctuations that seemed to follow the plane of the solar
system. Efstathiou said such anomalies might point to new frontiers
of physics, or at least to new mysteries. "I was explicitly told not
to say anything about God in this talk — which I've just violated,"
Efstathiou said.]

Fortunately or unfortunately the age at 13.81 Byrs can't be correct, or there can't be dark energy. I do agree 'new frontier or new mystery or 'god' is the Reality. Do you remember Prof. Fred Hoyle? One of the 'great' human beings- his thinking is close to the Reality after Newton during the past 300 years. And, the rest- the big-Bang experts and Einstein, Hubble etc… etc… missed the boat. If you want to know more, please write to me. I have a Book being published as we speak. By the way the 'god-like-particle' found by CERN and Prof. Incandela is a fiction.

Thanks,

Maheswar

A6

From: Mikki [mailto:stmmr@starpower.net]
Sent: Saturday, June 08, 2013 10:43 AM
To: 'correspondence_reply@durbin.senate.gov'
Cc: rao@tracobi.com; Congressman Elijah Cummings (md07ima@
mail.house.gov); Kleinman, Joan <Joan.Kleinman@mail.house.gov>
(Joan.Kleinman@mail.house.gov); FN-WHO-AAPI (AAPI@who.
eop.gov); 'ksamy@ceq.eop.gov'; 'Al Gore'; info@barackobama.
com
Subject: Climate change
06/08/2013

Dear Senator Durbin:

Thank you for your acknowledgment of my contact with your office after I met you briefly on June 4th in the Cannon Office Bldg. (at the affair of Hindu American Foundation).

At that time, I, also, met our good friend Congressman Cummings- I am expected to take an appointment to meet in his office for further discussion in regard to my research on the "cause for Rising Temperature on Earth". Also, on June 6th I briefly met Congressman Van Hollen (whom I consider as a young, bright leader).

I am hoping to meet with you, as the dynamic-leader in the Senate, and also work through Congressman Van Hollen and Congressman Cummings for help to put me in touch with the Whitehouse Science-Advisor. I am hoping to request the Whitehouse Science-Advisor to look it over my Research work, raise any questions/comments and if satisfied select 7 or more top-Scientists in the US (or for that matter from anywhere on Earth) to review my research work, and make a report to you, the two-Congressmen, and the President on their findings. That's all, a simple request. I will be happy to submit my 85-page research article to your office or to others, if so directed.

My associate Mr. Mandava, who lives in Orland Park near Chicago, will contact your local office very shortly; and, he will be happy to provide information on the topic of "Rising Temperature on Earth and the Primary-cause". Of course, the activity of Man has some contribution to it. I have been a good supporter of Al Gore, and I respect him greatly- but, unfortunately the confused-Scientists have mislead him on the cause of rising temperature.

On June 6th I met young-experts at the Whitehouse who work in the office of the Whitehouse Science-Advisor; I followed up with an e-mail message to them (see a copy, below).

I wish you, Congressman Van Hollen and Congressman Cummings would be kind enough to do the needful in the interest of all of us.

I am NOT asking for any Favor, I am only looking for the Experts to Review my work. The Experts are welcome to take me apart or agree with me- either way I will be happy!

Thanks,

Maheswar

From: Mikki
Sent: Friday, June 07, 2013 7:38 AM
To: FN-WHO-AAPI (AAPI@who.eop.gov); 'ksamy@ceq.eop.gov'
Subject: Climate change?
06/07/2013

Dear Gautam and friends:

I think the Whitehouse Briefing is excellent.

I enjoyed listening to the young-experts and able to communicate
on two issues that should interest all of us. And, I hope to continue
discussing with the experts for positive result.

(1) "Climate Change or Why the Rising Temperature?" Please
see the attached (a portion of a Book coming in July: Cover page,
preface & Veda Symbols- It is the Reality with which most, if not all
7 billion, People should be aware of [It is not just a Hindu concept-
it is the *Sanatana* Dharma (Eternal law of Nature, The Almighty)].
I am hoping even a 7th grader would be able to get it reading the
Book? The life on Earth will face rising temperature during the next
~200 years. Why? Look at the cover page: as our-Sun keeps orbiting
Indra (Neutron), it is going to catch up with its companion-Star (at
~130 AU apart). That means the space closer to our-Earth would
face 'heat' from our-Sun as well as from the companion-Sun. It is a
simple concept- so I need not explain. The question is- how come
our so called "great scientists" could not see that?. Sure, they knew
everywhere they look they see double Stars in space- except ours and
at another spot. They think, our-Sun is a loner and we have a 3-Star
system next to us. So, they goofed by lumping our-Companion
with another double-Star system and called it a 3-Star system. How
come? Their distance calculations in space are Wrong. They are,

310

after all, non-Engineers- never trained how to make measurements- that is the problem. I will be happy to submit my Research work (85-pages) to the Whitehouse Science-Advisor and I encourage any scientist to try and take it apart. I am sure I and my work will prevail.

(2) About the Patented "sleep method" I will directly contact Mr. Bhatia, soon.

Please review the attached portion of the Book, and feel free to write to me with any Qs or comments.

Thanks for your time, Maheswar

A8

From: Mikki
Sent: Tuesday, June 11, 2013 9:40 AM
To: 'IPCC-Sec@wmo.int'
Cc: rao@tracobi.com; Al Gore <algore@algore.com> (algore@algore.com)
Subject: Climate Change?
06/11/2013

Please forward this message to Dr. Pachauri.

From: Mikki [mailto:stmmr@starpower.net]
Sent: Tuesday, June 11, 2013 8:23 AM
To: FN-WHO-AAPI (AAPI@who.eop.gov); correspondence_reply@durbin.senate.gov; Kleinman, Joan <Joan.Kleinman@mail.house.gov> (Joan.Kleinman@mail.house.gov); Congressman Elijah Cummings (md07ima@mail.house.gov); Al Gore <algore@algore.com> (algore@algore.com); ksamy@ceq.eop.gov
Cc: rao@tracobi.com; stelthom@juno.com; Subra Baliga (subrabaliga@hotmail.com)
Subject: Climate Change?
06/11/2013

Dear Gautam:
Please see my response to NYT article by Gillis on rising temperature on Earth. Unfortunately, all of us including the scientists are lost: no-clue what is going on?

And, most of us refuse to understand the Reality, because each one of us Think we know it all !

The scientists have mislead a good man, Al Gore, and made him a spokesperson giving him a Nobel Prize- looks like the Devil is winning and God is losing? How long this status can continue- I guess until we see severe destruction during the next 100 or 200 years? Is there a Fix- may be not.

Please keep in touch, Maheswar

A9
[Attachment]
06/11/13
Dear Friend Gillis:

I read your article in NYT that "Global Temperatures Highest in 4,000 Years"- do you know Why?

I have a clear answer based on my research: The answer is, it is a ~1,000 year cycle. And, during ~250 years temperature on Earth will rise to be hot and ~250 years temperature will drop to be cold (or as the confused-scientists might say 'ice-age'). During the rest of the ~500 years, ~250 years temperature decreases from hot to warm, and ~250 years temperature increases from cold to warm (in other words we will have 2 transition periods of ~250 years each).

So, going back to your Heading- during the next ~200 years (if we are here) we will find 'highest' temperature over what the Earth has seen during the past 4,000 years. Why?

We have another 4.32×10^6 years Cycle (of which ~3.6×10^6 years have passed with ~700,000 years to go). During the remaining time of 700,000 years in this Cycle Earth will see the hot-temperature increasing in ~1,000 year cycle (for 250 years).

If you want to see my calculations, please write to me.

Maheswar

A10

From: Mikki
Sent: Sunday, July 28, 2013 5:28 AM
To: 'nicholas_devereux@warner.senate.gov'; 'mark_brunner@ warner.senate.gov'
Cc: Sen. Bill Nelson <bill@billnelson.senate.gov> (bill@billnelson. senate.gov); kevin_lefeber@durbin.senate.gov; Al Gore <algore@ algore.com> (algore@algore.com); FN-WHO-AAPI (AAPI@who. eop.gov); ksamy@ceq.eop.gov; 'SpeakerBoehner@mail.house. gov' (SpeakerBoehner@mail.house.gov); lars.heikensten@nobel.se; IPCC-Sec@wmo.int; 'scheduler@mikulski.senate.gov'
Subject: Atom (Indra)

07/28/13

Hi, Nicholas and Mark: [copy to President Obama at the Whitehouse with a request to forward it to his Science Advisor]

I met you and our good-Brother Sen. Warner in your office (at a coffee) in January 2011 and followed up: see my e-mail below at II. Also, I met Sen. Warner briefly in November 2012 and gave him the cover page of my Book (see attached. The Book will be out next month: god-Isvar…). Please note that my work involves the re-discovery of Reality- not a Philosophy or Religion or Fiction.

Also, I am pleased to submit to Sen. Warner a copy of my message sent to Sen. Cardin and Sen. Mikulski of MD and other leaders: see my e-mail below at I.

I am writing this message to Sen. Warner with a request to consider having a "Hearing" at the Senate Science Committee on the topic of my findings- especially, What is the primary cause for the Rising Temperature on Earth?

My finding discloses it is the "Nature" not the Man. Our good-leader Al Gore might disagree (since the scientists have misguided him…); but I am sure a good-person like Gore would listen with an open mind in the interest of all… The Nobel F and UNO's IPCC or any other Scientist is free to take me apart before the Senate Science Committee- I will be ready to tell you as it is.

I am hoping Sen. Nelson and Sen. Warner will lead to set a 'hearing' on the topic under the guidance of my leaders Sen. Cardin and Sen. Mikulski.

Please understand I am not asking you to do me a "favor"— do a "favor" to all of us !

Thanks.

Maheswar

A11
I From: Mikki
Sent: Saturday, July 27, 2013 1:54 PM
To: 'scheduler@mikulski.senate.gov'
Cc: Sen. Bill Nelson <bill@billnelson.senate.gov> (bill@billnelson.senate.gov); kevin_lefeber@durbin.senate.gov; Kleinman, Joan <Joan.Kleinman@mail.house.gov> (Joan.Kleinman@mail.house.gov); xan.fishman@mail.house.gov; 'Doug Walgren'
Subject: Thanks
07/27/13

Dear Senator:

I am a new kid (or an old kid of nearly 75) in MD.
I am a retired engineer (having run business for 30-years in Pittsburgh, Pa. and used to be a good-friend of late Sen. Heinz, and I am close-friend of ex-Congressman Doug Walgren there- he is now in DC area).

In the past 10-15 years I am out of loop with the leadership in DC (although recently I met and spoke to Sen. Durbin, Sen. Warner and few Congressman- like Chris Van Hollen....and I am in touch with John Delaney). I am out of loop because I have been doing Research in Cosmos from the point where Kepler, Galileo, Newton have left us 400 years ago- believe Einstein goofed and we are in a deep black-Hole. I can prove it; but, I am going thru some of the difficulties Galileo faced 400 years ago when he declared "Sun is the center of our universe, not the Earth".

My work tells me "*Indra* (Sanskrit word for Neutron) is the center of our universe, not the sun".

The present day scientists who worship god-Einstein and declare there is "no" other god (while wasting Billions on false research) would not care to even review my findings. That's the present status (although I am publishing a Book, soon).

Recently I met 2-staff members at Sen. Durbin's office. They have suggested that the Senate Science Committee Chaired by Bill Nelson might be the venue for a "hearing" where I can present my findings and any so called "top-scientists" can appear to take me apart. I love that chance- so, we can stop this Billions or Trillions of people's money from going waste on false research.

We can pick a simple topic for a short-Hearing: "What is the primary cause for the Rising Temperature on Earth? Is it the Nature or Man?"

I can prove to you, the Congress, and the scientists- it is the Nature. And this rise could continue for another 200 years before starting to decrease during the next 500 years- this cycle goes on in 1,000 year cycles until the End of present-Time in 700,000 years from now.

I hope I can find one Senator and one Representative to convince on this, and if convinced can take the leadership in this topic in the interest of all People in MD, America and on Earth.

Thank you.

Maheswar

A-12

II From: Mikki
Sent: Wednesday, January 26, 2011 12:21 PM
To: 'nicholas_devereux@warner.senate.gov'
Cc: 'mark_brunner@warner.senate.gov'....
Subject: Atom (Indra)
01/26/2011

Hi, Nicholas: *Namaste*!

Nice meeting you (introduced by Mr.Mark Brunner), and Neeta (introduced by Mr. Scott Price) at the Coffee this morning.

I am delighted to have met good-Senator Warner: when I said "... *India is the root of our Ancients... and my work shows we live in an Atom...*", Senator asked me "*is that on the basis of genetic study...*"

I said "based on '*Sanskrit*', our-root language, "*gods*", '*culture*' & '*features*'...(although I look darker than you... it is because I lived *too* long in the Sun); and I said "*I gave a brief description on a piece of paper to Neeta on my work... that should interest NASA*". Senator said "*I will read that...*"

I view of Senator's interest- I am happy to forward a copy of my recent e-mail sent to *Pope* Benedict on the Q of our *roots* with a '*Summary*' of my work Part I, for your kind information. I request you to pass it on to the Senator and others after you have had a chance to read it. I will be delighted to visit you to answer any Qs you or others might have on this...

Galileo (Kepler, Newton), 400 years ago informed us "*Earth is not, but Sun is the center of our-universe...*" After that, all Scientists including Einstein (and followers, like NASA), took that as the word of "*god*" or Bible and began *fishing in the desert* for knowledge like the *blind searching in the dark* (wasting time and Billions or Trillions…).

My work in Part I, following in the foot-steps of Kepler-Newton (plus the knowledge of our-Ancients in *Sanskrit* writings), shows ***"Sun is not, but Indra is the center of our-universe…"***

Now, on the Q of *"genetic study"* Senator raised- *genetics* is the study of *DNA* or *genes*; and *gene* is the sequence of *'atoms'* that control the *cell*- in you, I and the tree- so, we- all life on Earth- live in an Atom that creates us for a 'purpose'- whose Atom is this?

How did our-Ancients know to be able to inform us (thru *"Veda"* or knowledge of science in *Sanskrit*)?

Because, I am questioning NASA and the present crop of scientists- they ganged up against me, a little unknown *old-retired* engineer, hoping I will go away like Galileo- so they can keep wasting more Billions or Trillions of People's money as long as possible… (because they have no-*clue* as to the Reality).

Is there a way to Fix this? Why can't NASA pick 7 or more scientists with expertise in the field (if any still living…) and take a hard look at my work and allow me to make presentation and answer Qs…. This will take <u>no</u> more than a couple of days of time… to get it done… If they want Fee- maybe I can arrange it too…

Today, I am going to ask, again, NASA's director Bolden and Chief Scientist Hertz to see if are willing to re-consider my request… and, I will send you a copy to Senator Warner…

Thank you,

Maheswar

From: Mikki
Sent: Friday, January 07, 2011 12:59 PM
To: 'benedictxvi@vatican.va'
Cc: 'ohhdl@dalailama.com'; 'charles.bolden@nasa.gov'; 'Overbye, Dennis'; 'swh@damtp.cam.ac.uk'; 'naspresident@nas.edu'
Subject: Atom (Indra)

01/07/2011

Dear Pope Benedict: *Namaste* !

I think, being a German by birth, you are aware *Sanskrit* is the root language of all *Germanic-European* dialects, *Russian, Latin, Greek* etc.. including *English*- all written from <u>left to right</u>; *Avistan*, the Pharisee of Herat, Afghan, wrote *Sanskrit* in a different-script from <u>right to left</u> from which came the *middle-Eastern* dialects including *Aramaic* which *Jesus* spoke; and *Sanskrit* itself is rooted in (*Dravida*)-*Brahmi*, pictorial writing, from which came dialects in the *East*: Burma-Tibet to Australia-Indonesia-China to Japan-Americas... (Maya, Inca etc... are close to *Brahmi*'s Kannada, Telugu...). This is just a summary...

'*Veda*' in *Sanskrit* is knowledge of pure 'Science' known in yester-*Yuga* (4.32x10^6 years); and *Manu* must have figured out "*god*" or "*gods*" from this knowledge of Science... Thus, our-Culture of '*gods*' is rooted in *Sanskrit* too.

Now, the Q- how old is *Brahmi* or *Sanskrit*?

My calculations (based on IERS data on Earth wobble-spin) show we came out of Ice-age destruction ~25-30 thousand years ago (with Ice-age at Earth-spin of 22.5W to 22.5E: now, it is 23.5E). So, the Ice-age lasted for over ~**one million** years- that shows our-Ancients knew the knowledge of Science more than **1.6x10^6** years ago (or prior to the Ice-age destruction); so, *Brahmi* or *Sanskrit* must be much older than that. Your *Neanderthals* must have migrated into Europe before the Ice-age; and, the rest of your ancients must have migrated into Europe from Georgia (Eastern-Europe) after the Ice-age…., with some knowledge of Science and '*gods*' in *Sanskrit*. In ~150BC, Roman-army (per Julius Caesar's memoirs) did not know People live in that part of Germany until the villagers came out and fight bravely after worship of '*fire-god*'- one of the '*gods*' in *Veda*.

Now, we are in the process of *re*-discovering that knowledge. That re-discovery, really, began in 225BC with Archimedes' study in Alexandria, where Alexander the Great / later his followers found a library in the name of Alexander. In 325BC, Alexander took *Sanskrit* writings from Motherland *Bharat* (a portion of it is *India*-Alexander called '*Indoos*' for *Indus*-river). He, also, destroyed Pharisee's '*Zoroastrianism*' or '*Avistas*' (a misconstrued version of '*Veda*' in a <u>right to left</u> script, written after the Ice-age destruction- if it was before the Ice-age, Europe's *Bible* would be from <u>right to left</u> like *Quron* in middle-East). Then, came others- Italian-Galileo, German-Kepler, English-Newton, Russian-Mendeleev etc… Of course, we are aware Church-elitists persecuted Galileo because he propagated "*Earth is not, but Sun is the center of our-universe…*"

My work in the past >7 years that followed in the foot-steps of Kepler, Newton and Wien etc.., discovered that "*Sun is not, but **Indra** is the center of our-universe, and **Indra**-Suns-Planets together make up an Atom (Anuvu)…*" See the 'Abstract' attached… And, ask our-Noble Brother Dalai Lama who wrote a Book on Atom-*Indra* based on the knowledge of *Arya*-Brahmana or our-Druids.

Then the Q- whose Atom is this? According to our-Ancients, it belongs to Almighty-*Brahman*, our-*Father* (or *Mother*)- I am informed *Jesus* prayed to the *Father*...

Our scientists concede life on Earth is made of live-atoms, including the DNA; but, they are unable to comprehend we live within Almighty-*Brahman*. They worship Galileo eating Billions or Trillions of People's money, and tell us there is *"no-god"*. As an example take Prof. Steven Hawking- he recently wrote a book, and declared there is *"no god"*... In fact, I can prove Big bang, CMB, BH, Expanding U etc...or Darwin's evolution or finding of Man coming out of an Ape as the 'survivor of the fittest' is all a fiction, and the reality is quite different...

Please feel free to contact me if any Qs... on this.

Respectfully,

Maheswar

God was behind Big Bang, pope says: 'The universe is not the result of chance, as some would want to make us believe'

VATICAN CITY — God's mind was behind complex scientific theories such as the Big Bang, and Christians should reject the idea that the universe came into being by accident, Pope Benedict said Thursday.

> "The universe is not the result of chance, as some would want to make us believe," Benedict said on the day Christians mark the Epiphany, the day the Bible says the three kings reached the site where Jesus was born by following a star.
>
> "Contemplating it (the universe) we are invited to read something profound into it: the wisdom of the creator, the

inexhaustible creativity of God," he said in a sermon to some 10,000 people in St. Peter's Basilica on the feast day.

While the pope has spoken before about evolution, he has rarely delved back in time to discuss specific concepts such as the Big Bang, which scientists believe led to the formation of the universe some 13.7 billion years ago.

Researchers at CERN, the nuclear research center in Geneva, have been smashing protons together at near the speed of light to simulate conditions that they believe brought into existence the primordial universe from which stars, planets and life on earth — and perhaps elsewhere — eventually emerged.

Proof God doesn't exist?
Some atheists say science can prove that God does not exist, but Benedict said that some scientific theories were "mind limiting" because "they only arrive at a certain point ... and do not manage to explain the ultimate sense of reality."

He said scientific theories on the origin and development of the universe and humans, while not in conflict with faith, left many questions unanswered.

"In the beauty of the world, in its mystery, in its greatness and in its rationality ... we can only let ourselves be guided toward God, creator of heaven and earth," he said.

Benedict and his predecessor John Paul have been trying to shed the Church's image of being anti-science, a label that stuck when it condemned Galileo for teaching that the earth revolves around the sun, challenging the words of the Bible.

Galileo was rehabilitated and the Church now also accepts evolution as a scientific theory and sees no reason why God could not have used a natural evolutionary process in the forming of the human species.

The Catholic Church no longer teaches creationism — the belief that God created the world in six days as described in the Bible — and says that the account in the book of Genesis is an allegory for the way God created the world.

But it objects to using evolution to back an atheist philosophy that denies God's existence or any divine role in creation. It also objects to using Genesis as a scientific text.

A13

From: Mikki
Sent: Tuesday, July 30, 2013 2:25 PM
To: Sen. Bill Nelson <bill@billnelson.senate.gov> (bill@billnelson.senate.gov); nicholas_devereux@warner.senate.gov; mark_brunner@warner.senate.gov; kevin_lefeber@durbin.senate.gov; 'scheduler@mikulski.senate.gov'
Cc: xan.fishman@mail.house.gov; Kleinman, Joan <Joan.Kleinman@mail.house.gov> (Joan.Kleinman@mail.house.gov); carla.coleman@mail.house.gov; lars.heikensten@nobel.se; IPCC-Sec@wmo.int; Al Gore <algore@algore.com> (algore@algore.com); 'Doug Walgren'
Subject: Model for Rising Temp. on Earth?
07/30/13

Dear Senators: Nelson, Warner, Durbin, Mikulski and Cardin:

Please review the e-mail message below- the latest that shows the rising CO2 has nothing to do with the rising temperature on the earth. As I said before, I can prove that to your satisfaction.
We get this weather change in ~1,000 year cycles and next ~200 years the temperature goes up (or the next ~700,000 years until the End of Time). Please do not be scared. This has been happening for many millions of years in a 4.32 million year cycles. We, the life, will come and go. Some of us do 'god's work' and not come back while others keep coming back until they finish their time… Our 'godly' ancients knew all about it, and informed us. We did not listen- most of us (not all), being ignorant.

I hope you will take up your Righteous duty and try to Fix it. Please contact me if I can assist.

Maheswar

A14

From: Mikki
Sent: Tuesday, July 30, 2013 1:35 PM
To: 'ashley.ballantyne@umontana.edu'
Cc: robert duncan (info@eurasiareview.com)
Subject: Model for Rising Temp. on Earth?

07/30/13

Dear Prof. Ballantyne:

I read an article in Today's "Eurasia Review" which is based on your research at University of Colorado, Boulder, on the topic "Ice-Free Arctic Winters Could Explain Amplified Warming During Pliocene"- I cite the full article below.

I tried to reach you this morning and left a message. Also, I tried to find a contact to Prof. Jim White at UC-Institute of Arctic and Alpine Research at Boulder, and not successful, yet- U of Colorado Webpage and Telephones are down (under repair). Please forward a copy of my e-mail to Prof. White with a request to contact me, directly. I like your work (although you will not get there until you read and understand my work). I have some direct interest in your findings such as:

(1) Taking as correct that "***In early May 2013,.....the concentration of carbon dioxide climbed to 400 parts per million for the first time in modern history***...", please tell me on what basis you believe or estimate that "***The last time... the carbon dioxide concentration in the atmosphere reached 400 ppm—between 3 and 5 million years ago during the Pliocene***—the ***Earth was about 3.5 to 9 degrees Fahrenheit warmer ...than it is today***. During that time period, trees overtook the tundra, sprouting right to the edges

327

of the Arctic Ocean, and the seas swelled, pushing ocean levels 65 to 80 feet higher."? That is a fundamental Q with which you must be sure of before going forward. Please see my comment below:

(2) I agree with Prof. White when he says that *"When we put 400 ppm carbon dioxide into a model, we don't get as warm a planet as we see when we look at paleorecords from the Pliocene. That tells us that there may be something missing in the climate models."* <u>My comment</u>: Use of the 'word paleorecods' is wrong if the Prof. claims that record is 3-5 million years old. The record he saw changes in ~1,000 year cycles- that is what is missing in the climate models. You have to read my work to understand that. And the 3-5 million years is actually 4.32 million years that takes major changes with the End of Life on Earth as we know it. Also, <u>I know the higher levels of CO2 has nothing to do with the Rising Temperature on Earth</u>.

(3) I, also, agree with Prof. White when he says that *"We tried a simple experiment in which we said, 'We don't know why sea ice might be gone all year round, but let's just make it go away.' And what we found was that we got the right kind of temperature change and we got a dampened seasonal cycle, both of which are things we think we see in the Pliocene."* <u>My comment</u>: The ice goes away in the hemisphere that faces the sun (which we call it Arctic, now) in a 1,000 year cycle or in the next 200 years.

(4) I do not agree with Prof. White when he says that *"Basically, when you take away the sea ice, the Arctic Ocean responds*

by creating a blanket of water vapor and clouds that keeps the Arctic warmer." My comment: A wrong assumption. Water vapor and clouds has nothing to do to keep the Arctic warmer….

(5) Prof. White also says "*We're trying to understand what happened in the past but with a very keen eye to the future and the present. The piece that we're looking at in the future is what is going to happen as the Arctic Ocean warms up and becomes more ice-free in the summertime. Will we continue to return to an ice-covered Arctic in the wintertime? Or will we start to see some of the feedbacks that now aren't very well represented in our climate models? If we do, that's a big game changer.*" I agree, it is a good thinking- but, the Professor would miss the boat like so many scientists have in the past, including Einstein, the great. Please see my conclusion or comment below:

My comment: Please take a note of another good article in Eurasia Review dated Nov. 12th 2012 (and NYT) from which I quote here-

By: <u>Eurasia Review</u> November 12, 2012

Don't graphs show that current temperatures are the highest in 1,000 years?

Penn State professor and UN IPCC modeler Michael Mann did publish a hockey stick-shaped graph that purportedly showed an unprecedented sudden increase in average global temperatures, following ten centuries of supposedly stable climate. However, Dr. Mann was at the center of the Climate-gate scandal. His graph and the data and methodology behind it have been scrutinized and debunked in peer-reviewed studies by numerous climate scientists, statisticians and other experts.

The latest research clearly reveals that the Medieval Warm Period (also called the Medieval Climate Optimum) has been verified and was in fact global, not just confined to the Northern Hemisphere. The Center for the Study of Carbon Dioxide and Global Change reported in 2009 that "the Medieval Warm Period was: (1) global in extent, (2) at least as warm as, but likely even warmer than, the Current Warm Period, and (3) of a duration significantly longer than that of the Current Warm Period to date."

The Science and Public Policy Institute reported in May 2009: "More than 700 scientists from 400 institutions in 40 countries have contributed peer-reviewed papers providing evidence that the Medieval Warm Period (MWP) was real, global, and warmer than the present. And the numbers grow larger daily."

New York Times- January 09, 2013
"Those of us who spend our days trawling - and contributing to - the scientific literature on climate change are becoming increasingly gloomy about the future of human civilization," Dr. Elizabeth Hanna,

a researcher at the Australian National University in Canberra, told The Sydney Morning Herald. "We are well past the time of niceties, of avoiding the dire nature of what is unfolding, and politely trying not to scare the public."

My research concludes Man is Not primarily responsible. Nature is. And, the temperature gets hotter during 250 years in every 1,000 year cycle. The summary of the above 2-News items informs us:

The temperature was hotter during MWP or 1,000 years ago, with millions of people without industry or CO_2; now, the temperature is getting hotter (not as hot as in MWP) despite billions of people with industry and CO_2. Therefore, what is the reason or basis for putting all the blame on the Man for the 'Climate Change'?

So, dear Professors Ballantyne and White here is the truth- are you ready? Accept these changes occur every 1,000 years and concentrate your work or model to predict in ~1,000 year cycles. To handle in 4.32 million year cycles is beyond your or my understanding of the Nature. Man is Nothing when compared to Almighty-god. If you want to review my work, please write to me- I will send you the article (85 pages) with a condition that you agree to make comments- both good and bad if any.

Thanks,
Maheswar

By Eurasia Review July 29, 2013

> Year-round ice-free conditions across the surface of the Arctic Ocean could explain why the Earth was substantially warmer during the Pliocene Epoch than it is today, despite similar concentrations of carbon dioxide in the atmosphere, according to new research carried out at the University of Colorado Boulder.

In early May, instruments at the Mauna Loa Observatory in Hawaii marked a new record: The concentration of carbon dioxide climbed to 400 parts per million for the first time in modern history.

The last time researchers believe the carbon dioxide concentration in the atmosphere reached 400 ppm—between 3 and 5 million years ago during the Pliocene—the Earth was about 3.5 to 9 degrees Fahrenheit warmer (2 to 5 degrees Celsius) than it is today. During that time period, trees overtook the tundra, sprouting right to the edges of the Arctic Ocean, and the seas swelled, pushing ocean levels 65 to 80 feet higher.

Scientists' understanding of the climate during the Pliocene has largely been pieced together from fossil records preserved in sediments deposited beneath lakes and on the ocean floor.

"When we put 400 ppm carbon dioxide into a model, we don't get as warm a planet as we see when we look at paleorecords from the Pliocene," said Jim White, director of CU-Boulder's Institute of Arctic and Alpine Research and co-author of the new study published online in the journal Palaeogeography, Paleoclimatology, Palaeoecology. "That tells us that there may be something missing in the climate models."

Scientists have proposed several hypotheses in the past to explain the warmer Pliocene climate. One idea, for example, was that the formation of the Isthmus of Panama, the narrow strip of land linking North and South America, could have altered ocean circulations during the Pliocene, forcing warmer waters toward the Arctic. But many of those hypotheses, including the Panama possibility, have not proved viable.

For the new study, led by Ashley Ballantyne, a former CU-Boulder doctoral student who is now an assistant professor of bioclimatology at the University of Montana, the research team decided to see what would happen if they forced the model to assume that the Arctic was free of ice in the winter as well as the summer during the Pliocene. Without these additional parameters, climate models set to emulate atmospheric conditions during the Pliocene show ice-free summers followed by a layer of ice reforming during the sunless winters.

"We tried a simple experiment in which we said, 'We don't know why sea ice might be gone all year round, but let's just make it go away,' " said White, who also is a professor of geological sciences. "And what we found was that we got the right kind of temperature change and we got a dampened seasonal cycle, both of which are things we think we see in the Pliocene."

In the model simulation, year-round ice-free conditions caused warmer conditions in the Arctic because the open water surface allowed for evaporation. Evaporation requires energy, and the water vapor then stored that energy as heat in the atmosphere. The water vapor also created clouds, which trapped heat near the planet's surface.

"Basically, when you take away the sea ice, the Arctic Ocean responds by creating a blanket of water vapor and clouds that keeps the Arctic warmer," White said.

White and his colleagues are now trying to understand what types of conditions could bridge the standard model simulations with the simulations in which ice-free conditions in the Arctic are imposed. If they're successful, computer models would be able to model the transition between a time when ice reformed in the winter to a time when the ocean remained devoid of ice throughout the year.

Such a model also would offer insight into what could happen in our future. Currently, about 70 percent of sea ice disappears during the summertime before reforming in the winter.

"We're trying to understand what happened in the past but with a very keen eye to the future and the present," White said. "The piece that we're looking at in the future is what is going to happen as the Arctic Ocean warms up and becomes more ice-free in the summertime.

"Will we continue to return to an ice-covered Arctic in the wintertime? Or will we start to see some of the feedbacks that now aren't very well represented in our climate models? If we do, that's a big game changer."

A15

From: Mikki
Sent: Saturday, August 03, 2013 8:05 AM
To: Sen. Bill Nelson <bill@billnelson.senate.gov> (bill@billnelson.senate.gov)
Cc: nicholas_devereux@warner.senate.gov; kevin_lefeber@durbin.senate.gov; 'scheduler@mikulski.senate.gov'; xan.fishman@mail.house.gov; Kleinman, Joan <Joan.Kleinman@mail.house.gov> (Joan.Kleinman@mail.house.gov); 'SpeakerBoehner@mail.house.gov' (SpeakerBoehner@mail.house.gov); charles.bolden@nasa.gov; lars.heikensten@nobel.se; 'Doug Walgren'; ksamy@ceq.eop.gov; FN-WHO-AAPI (AAPI@who.eop.gov); Al Gore <algore@algore.com> (algore@algore.com); carla.coleman@mail.house.gov
Subject: Van Allen radiation belts?
08/03/13

Dear Sen. Nelson:

Please see the latest confusion of our present day scientists at the expense of People's money. There is not much I can say other than 'bring them all into a room' and find out what is going on?
They think they know it all although confused- but, not willing to listen to reason. That is the state of affairs- we are going out of control acting like Kings in wasting People's money?

Maheswar

From: Mikki [mailto:stmmr@starpower.net]
Sent: Saturday, August 03, 2013 7:18 AM
To: reeves@lanl.gov
Cc: 'science_editors@aaas.org'
Subject: Van Allen radiation belts?
08/03/13

Dear Dr. Reeves:

Regarding your article in Science Journal of July 25th 2013 on the topic **"Electron Acceleration in the Heart of the Van Allen Radiation Belts"**, I did not hear from you, again- in response to my inquiry below.

I wish you sent me a link at DOE's Los Alamos National Lab. or some other source where I could review your Figures, Data, and the Findings and Help you out of the confusion.
I mean I know what is going on and I could have assisted you and rest of the scientists from the confused state of the matter, and put to bed on the Q of what actually is going on between the earth and sun.

I hope you are trying to figure that out from NASA data and the hints I gave you. I assure, that will not work because there is more to it than you can imagine or able to see from NASA data.

As I said before, the data you think you have "tonnes more exciting stuff to look at with the Van Allen Probes data, some of which will be things that the satellites were designed to look for and some completely unexpected…" would be of NO USE to figure out what is going on- I truly believe you must read my Book "god-Isvar, Swastika-science" which will be out this month to see head and tail of what is going on in Cosmos.

I can give you another hint: Newton simply said 'gravity', but did not know much nor explain the mechanism of transfer of the forces between Earth (or the planets) and the sun- he had NO Clue.
Your so called 'Van Allen Belt' is small, very insignificant, portion of the mechanism that is in play, here- actually, it is wrong to call it 'Van Allen Belt' because he nor any other scientist knew what is going on.

See your own Abstract of July 25th 2013 below to prove my point:

"The Van Allen Radiation Belts contain ultra-relativistic electrons trapped in the Earth's magnetic field. Since their discovery in 1958, a fundamental unanswered question has been how electrons can be accelerated to such high energies. Two classes of processes have been proposed: (i) transport and acceleration of electrons from a source population located outside the radiation belts (radial acceleration); or (ii) acceleration of lower-energy electrons to relativistic energies in situ, in the heart of the radiation belts (local acceleration). We report measurements from NASA's Van Allen Radiation Belt Storm Probes that clearly distinguish between the two types of acceleration. The observed radial profiles of phase space density are characteristic of local acceleration in the heart of the radiation belts and are inconsistent with a predominantly radial acceleration process."

Also, see portions of the article in the 'Physics World.Com' and your comments:

The article says "The Van Allen radiation belts are two concentric, doughnut-shaped rings that are made up of high-energy electrons that vary in intensity. The belts are confined within the Earth's magnetosphere and extend from about 1000 to 60,000 km above the Earth's surface. Although discovered by <u>American physicist James Van Allen</u> more than 50 years ago, the Van Allen belts are not yet fully understood. According to the lead researcher of the new study, <u>Geoffrey Reeves</u> at the Los Alamos National Laboratory, New Mexico in the US, previous observational data taken in the 1990s did not fit the conventional theories of the time, leading researchers to question and debate over what processes really control the intensity of the radiation belts... Reeves told *physicsworld.com* that previous satellite missions had provided tantalizing evidence for local acceleration, but there were always limitations. 'Essentially, neither side of the argument could convince the other side that local acceleration did or did not happen'. He goes on to explain that Van Allen Probes have three features that make the latest observations unique, including 'the right instruments spanning a broad range of energies with amazing sensitivity, an equatorial orbit that cuts through the belts at different altitudes and two satellites that can unambiguously resolve whether something is changing in time or in space or both."

So, I say this- First, the concept of "doughnut-shaped rings" is Wrong and your data may be partially correct, but you or any of the present day scientists will never get it- because the thinking is flawed.

Sorry to be blunt- I have been telling this for more than 5-years to the so called 'top-scientists', 'editors of Science Journals... Who cares- they have Billions or Trillions coming freely why not keep doing Fiction as long as possible? That is the summary in essence. Even the Nobel and the members of Congress seem to be slow in Fixing the problem- but, how long can it continue?

Please feel free to write to me if I can help.

Thanks, Maheswar

From: Mikki
Sent: Thursday, August 01, 2013 8:15 PM
To: 'Reeves, Geoffrey D'
Subject: Van Allen

Thank you Sir, for your response- no I could not open- they want me to purchase at a price if not a member.

If you do not have a copy of that same article- I can read something close to it that you might have on your Webpage- could you send me the link to your Webpage or DOE site where I can read and review the observed data?

Let me share what I think is going on: yes, Earth's magnetic field lines go from N-magnetic pole towards S-magnetic pole and the sun's magnetic field lines do the same at much larger speeds while hitting Earth's lines at an angle ~70 degrees to Earth's axis of rotation- I am sure your instruments, I think, are not yet capable to read the data that precisely. That might be the reason your data is not clear and you think the flows are going both radial and parallel? Before I conclusively assert, I must review the various Figs you have prepared using the data you have....

I have more to say- but, I will stop and wait to read on your findings and the supporting data.

Thanks, Maheswar

From: Reeves, Geoffrey D [mailto:reeves@lanl.gov]
Sent: Thursday, August 01, 2013 5:30 PM
To: Mikki
Subject: Van Allen

I would be happy to have you read the paper. Science told me that this link should be open access to the PDF
http://www.sciencemag.org/content/early/2013/07/24/science.1237743.full.pdf?keytype=ref&siteid=sci&ijkey=M3pdEmBEDwMo6

Please let me know if you have any trouble with the link.

Thanks

Geoff

On Aug 1, 2013, at 3:02 PM, Mikki wrote:

08/01/13

Dear Prof. Reeves:

I read an article in Physics World on your findings "**Electron Acceleration in the Heart of the Van Allen Radiation Belts**"

I think, I know what is going on in this process- before I write to you about it, I would like to read carefully your entire article- then, I can make my comments and send it to you.

Would you kindly forward that article to me? I am an old retired engineer near Washington, DC (not close to a science library or I am not a member of Science Journal).

Thanks, Maheswar

Geoff Reeves	ph 505-665-3877
Space Science and Applications Group	fax 505-665-7395
Los Alamos National Laboratory	
Los Alamos, NM 87544	sec 505-667-2701
http://www.rbsp-ect.lanl.gov/people/reeves/	

A16

From: Mikki
Sent: Monday, August 05, 2013 7:10 PM
To: 'Reeves, Geoffrey D'
Subject: Van Allen radiation belts?
08/05/13

Thank you Sir:
Today I went to Georgetown U Lib. and copied your paper, Kerr's
paper & Baker's paper from Science.

I just now came back from DC- I will read them tonight or at least
try to get a gist and see where you are all heading with this?

Please, believe me there is not much to be confused and it may look
'different facets of this vast Cosmos'- but, it is not really.

We are all talking about the same Reality. If the publisher would
allow me to add few sentences to clarify what I have already set out
in the Book (not my 85-page research paper which deals with the
science part meant to the Scientist)- the Book explains the Reality
to a scientist as well as to a lay person- then, that would make more
sense to you, and hopefully, you and I convince rest of the World?

I believe a smart person, like you, trying to understand the riddle
Kepler-Newton tried to solve (but, could not- came close to 75%...)
plus the clues left by Van Allen (unknowingly...) should make it
clear when you review my Book. Let us keep in touch.

Maheswar

From: Reeves, Geoffrey D [mailto:reeves@lanl.gov]
Sent: Monday, August 05, 2013 12:30 PM
To: Mikki
Subject: Re: Van Allen radiation belts?

Maheswar....

I am sorry I didn't respond quickly enough to your request. I was out of the office most of Friday and e-mail does tend to pile up. I've attached a copy of our paper and the supplemental material. I would be happy to take a look at your book once it is published but it sounds, perhaps, that we are studying different facets of this vast cosmos we live in.

best regards

Geoff

A17

From: Mikki

Sent: Sunday, August 18, 2013 10:35 PM

To: xan.fishman@mail.house.gov; karen_murphy@mikulski.senate.gov; royce_brooks@cardin.senate.gov; kevin_lefeber@durbin.senate.gov; nicholas_devereux@warner.senate.gov; Sen. Bill Nelson <bill@billnelson.senate.gov> (bill@billnelson.senate.gov); carla.coleman@mail.house.gov

Cc: reeves@lanl.gov; 'jwhite@colorado.edu'; IPCC-Sec@wmo.int; lars.heikensten@nobel.se; Al Gore <algore@algore.com> (algore@algore.com); FN-WHO-AAPI (AAPI@who.eop.gov); ksamy@ceq.eop.gov; overbye@nytimes.com; 'incandel@hep.ucsb.edu'; Kleinman, Joan <Joan.Kleinman@mail.house.gov> (Joan.Kleinman@mail.house.gov); charles.bolden@nasa.gov

Subject: Rising Temp. on Earth & Van Allen radiation belts? 08/18/13

Although my work deals with a number of major issues that are clearly discussed in the Book "god-Isvar, Swastika-science" which will be out this month- I wish to leave, here, couple of simple concepts for your kind consideration:

(1)   Did CERN see a 'god-particle' at $125 \times 10^9 eV$? I think, not- Man is not ready to see or sense that, yet.

(2)   Did NASA/Reeves team see 'gravity' in Van Allen radiation belts which Newton missed? I hope so- see the e-mails below.

(3)   Did IPCC Chair Pachauri/Gore see Man do it (cause the Temp. to rise on Earth)? I hope, not. Why not? Simple: The concept of covering a 'living-body' with a 'thicker blanket' (like a 'thicker greenhouse' around Earth with rising CO_2 caused by Man) to insulate, at least partially, the 'radiating' body-heat from going

out or the 'cold' outside from coming in during the winter works just fine in maintaining the body at 98.4F.

But, the same blanket would not do any good during the summer – in fact, the blanket makes the 'living-body' feel 'hotter'- Why?

It is not the atoms (or CO_2) in the blanket that decided to create the heat by suddenly starting to 'split' or 'fuse'- it is the change "outside" from winter to summer.

Therefore, the 'change' must be taking place "outside" of the blanket and the "greenhouse" has nothing to do with the 'rising temperature'.

So, you may ask- Why this Change Outside of the Greenhouse?

You will find the correct answer in my Book- I can't do justice in trying to explain in few words…

I hope to see you, soon, and answer your Q.

Thanks,

Maheswar

From: Mikki
Sent: Friday, August 09, 2013 9:49 PM
To: 'Reeves, Geoffrey D'
Subject: Van Allen radiation belts?
08/09/13

Dr. Reeves:

Thanks for the article (including the supplement…). I read it quickly. I will read it, again-later.

I agree the observed data tells us 'there is something cooking', but it will not tell us what is cooking until we are able to see the Big-picture. I mean is this thing cooking 'gravity'?
I think, yes.

By the way, these are not the trapped electrons or protons accelerating in 'local' and 'radial' directions (as you put it).

These are the 'atoms'- in Sanskrit it is 'paramanuvu', atoms in infinite size.

As you know, atom consists of electron, proton, and neutron- recall atom is 'indivisible'? That is true.

Proton or electron can't exist in the absence of neutron, the master, creator / destroyer of proton / electron.

In Reality, our-Sun and the planets (Earth included) are created by Neutron (in Sanskrit it is 'Indra').

Indra creates the suns, planets, you, me and the tree…. and destroys- all for a purpose. The Book coming this month explains all that (based on my >10 years of research from the point where Kepler, Newton have left us).

As I indicated to you in my previous e-mail, the 'local' and 'radial' accelerations, and your 'phase space density' etc… is real and being created by three actors: Indra, sun and earth.

They, while acting at different angles, create 'moving force-bands' between Indra-sun, Indra-earth & sun-earth: you call these force-bands a magnetic field where the neutrons having protons at different temperatures 'fuse' and 'split' as the actors keep moving…. So, what you have observed are the acts- when you observe the atoms within the force-band it appears they are trapped; and when you observe outside of the force-band it appears they are accelerating (at varied phase space density) in local and radial directions due to the 'fusing and splitting' of the atoms within the force-bands as a result its moving action created / destroyed by these three actors.

This is what actually going on in Cosmos and that explains the puzzle Newton could not explain and simply left calling it as 'gravity'.

Please feel free to agree or disagree with a comment…

Maheswar

From: Reeves, Geoffrey D [mailto:reeves@lanl.gov]
Sent: Monday, August 05, 2013 12:30 PM
To: Mikki
Subject: Re: Van Allen radiation belts?

Maheswar,

I am sorry I didn't respond quickly enough to your request. I was out of the office most of Friday and e-mail does tend to pile up. I've attached a copy of our paper and the supplemental material. I would be happy to take a look at your book once it is published but it sounds, perhaps, that we are studying different facets of this vast cosmos we live in.

best regards

Geoff

A18

From: Mikki
Sent: Wednesday, August 21, 2013 8:58 AM
To: IPCC-Sec@wmo.int; Al Gore <algore@algore.com> (algore@
algore.com); lars.heikensten@nobel.se
Cc: Sen. Bill Nelson <bill@billnelson.senate.gov> (bill@billnelson.
senate.gov); nicholas_devereux@warner.senate.gov; kevin_lefeber@
durbin.senate.gov; karen_murphy@mikulski.senate.gov; Kleinman,
Joan <Joan.Kleinman@mail.house.gov> (Joan.Kleinman@mail.
house.gov); xan.fishman@mail.house.gov; carla.coleman@mail.
house.gov; charles.bolden@nasa.gov; royce_brooks@cardin.senate.
gov; FN-WHO-AAPI (AAPI@who.eop.gov); overbye@nytimes.
com; reeves@lanl.gov; 'jwhite@colorado.edu'; ksamy@ceq.eop.gov
Subject: Rising Temp.?

08/21/13

Dear Dr. Pachauri and Nobel Gore:

On 08/18/13, I sent you my message (see below) and on 08/19/13,
your leaked report assures us with '95% certainty' that man is the
principal cause for the rising temperature on the earth, melting
ice and rising sea levels and other coming destructions during the
next few hundred years- because man is dumping CO_2 into the
atmosphere.

I want you to know, I am with you on the issue of 'man dumping
CO_2' as if there is no tomorrow- that act must be curtailed or man
would become responsible for the injury to the health/life of the
species.

Why not tell the truth 'as is' and help reduce CO_2 and other garbage
man is creating?

Instead, you are creating a 'religion' of your own mixing apples and oranges to 'divide' man- like 'they' did it with Avesta, Torah, Bible, Koran, Red book and with the Bombs etc. to Win- for What purpose?

I agree with your goal to do it right- but disagree with your methods. You are wrong to advocate with '95% certainty' that man is the principal cause for rising temperature on the earth etc.

I am ready and willing to debate with you or the 'thousands' of your scientists on this one issue- Whether the nature is the cause or man?

If you are ready to do it right, please delete that '95% certainty' from your report and ask me to submit my work for your review- just to see if it makes sense?

Thank you,
Maheswar

From: Mikki
Sent: Sunday, August 18, 2013 10:35 PM
To: xan.fishman@mail.house.gov; karen_murphy@mikulski.senate.gov; royce_brooks@cardin.senate.gov; kevin_lefeber@durbin.senate.gov; nicholas_devereux@warner.senate.gov; Sen. Bill Nelson <bill@billnelson.senate.gov> (bill@billnelson.senate.gov); carla.coleman@mail.house.gov
Cc: reeves@lanl.gov; 'jwhite@colorado.edu'; IPCC-Sec@wmo.int; lars.heikensten@nobel.se; Al Gore <algore@algore.com> (algore@algore.com); FN-WHO-AAPI (AAPI@who.eop.gov); ksamy@ceq.eop.gov; overbye@nytimes.com; 'incandel@hep.ucsb.edu'; Kleinman, Joan <Joan.Kleinman@mail.house.gov> (Joan.

Kleinman@mail.house.gov); charles.bolden@nasa.gov
Subject: Rising Temp. on Earth & Van Allen radiation belts?

08/18/13

Although my work deals with a number of major issues that are clearly discussed in the Book "god-Isvar, Swastika-science" which will be out this month- I wish to leave, here, couple of simple concepts for your kind consideration:

(1) Did CERN see a 'god-particle' at $125 \times 10^9 eV$? I think, not- Man is not ready to see or sense that, yet.

(2) Did NASA/Reeves team see 'gravity' in Van Allen radiation belts which Newton missed? I hope so- see the e-mails below.

(3) Did IPCC Chair Pachauri/Gore see Man do it (cause the Temp. to rise on Earth)? I hope, not. Why not?

Simple: The concept of covering a 'living-body' with a 'thicker blanket' (like a 'thicker greenhouse' around Earth with rising CO_2 caused by Man) to insulate, at least partially, the 'radiating' body-heat from going out or the 'cold' outside from coming in during the winter works just fine in maintaining the body at 98.4F.

But, the same blanket would not do any good during the summer - in fact, the blanket makes the 'living-body' feel 'hotter'- Why?

It is not the atoms (or CO_2) in the blanket that decided to create the heat by suddenly starting to 'split' or 'fuse'- it is the change "outside" from winter to summer.

Therefore, the 'change' must be taking place "outside" of the blanket and the "greenhouse" has nothing to do with the 'rising temperature'.

So, you may ask- Why this Change Outside of the Greenhouse?

You will find the correct answer in my Book- I can't do justice in trying to explain in few words.

I hope to see you, soon, and answer your Q.

Thanks,
Maheswar

A19

From: Mikki
Sent: Monday, August 26, 2013 10:00 AM
To: xan.fishman@mail.house.gov; karen_murphy@mikulski.senate.
gov; royce_brooks@cardin.senate.gov
Cc: Sen. Bill Nelson <bill@billnelson.senate.gov> (bill@billnelson.
senate.gov); nicholas_devereux@warner.senate.gov; kevin_lefeber@
durbin.senate.gov; Kleinman, Joan <Joan.Kleinman@mail.house.
gov> (Joan.Kleinman@mail.house.gov); carla.coleman@mail.house.
gov; Al Gore <algore@algore.com> (algore@algore.com); lars.
heikensten@nobel.se
Subject: rising temp.?

08/26/13

On Thursday I left a message to Ms. Royce Brooks if we can meet
this week. I am sure she was busy on Friday.

The only purpose in my trying to reach out to your office is-
(1)   To share my findings on the 'primary cause for the rising
temperature' with you and answer Qs- because it is my 'duty' to tell
the 'truth' to whoever wants to listen.
That includes the 'confused-scientists. See the e-mail to a scientist at
Max Planck Institute, below.
(2)   To find out if the Science Committee in the Congress would
have any interest to bring the so called 'top scientists' and 'me' into a
Room and find out what is really going on?
Please don't mind my 'blunt or straight talk'- I feel 'bad' to see my
Brothers go in a wrong path misled by the 'confused-scientists' and
walk straight into a 'deep black-Hole' the great Einstein devised to
beat our 'godly' ancients- 'Avesta' of the great Zoroaster of Afghan
did the same to Man for (tens of) thousands of years....

We must move forward or expect more destruction as a result of divisions and the nature….

Maheswar

From: Mikki
Sent: Saturday, August 24, 2013 8:52 AM
To: 'markus.reichstein@bgc-jena.mpg.de'
Subject: CO2

08/24/13

Dear Dr. Reichstein:

I read a news item this morning on your findings from a model study on 'climate change' due to the man-made CO2 in the atmosphere-the cause for the rising temperature, droughts etc... on the earth.
I will read your article which I understand you have posted in your Website at Max Planck Biogeochemistry Institute and I try to learn more details.

I am a retired old engineer and reside near Washington, DC, USA. I have been involved in the study-research of Cosmos for more than 10-years.
I began my work from the point where Kepler, Newton left us some 300-400 years ago.
I find increase in CO2 in the 'greenhouse gasses' or the 'blanket around Earth' is not the cause for rising temperature on the earth or melting Ice and rising Sea levels or droughts or climate change...
Then, you may ask- what is the cause? My answer: primarily the nature- how come?

Sun orbits around '*Indra*' (a *Sanskrit* word for Neutron), like Earth orbits around Sun. The only difference being Sun's orbit takes ~700 years. During this time, Sun comes closer to a Star (a companion) once in every ~1,000 years. So, during this 1,000 year cycle, about ¼ of the time or 250 years Earth feels warmer to hotter climate- like the N. Hemisphere on Earth feeling summer in June-September each year in a 12-month cycle...

My finding is correct. You can check and find why.

Take a realistic example: In winter you cover with a 'blanket' or a thicker dress to keep yourself 'warm'; in summer if you keep the same 'blanket' you feel 'hot' with rising temperature around your body- correct?

Why? You made no change. Yet, the climate around your body feels a change- correct?

What is the root cause for this change? Winter v. Summer.

You or I have no role to change winter into summer? So, is the earth- it simply keep on radiating some heat like you or I do....

In the same way, Earth (plus you or I) has no role to change the temperature around its own body- the 'blanket of greenhouse gasses around Earth' does a similar task like the blanket or the dress you or I put on the body.

Of course, thicker the dress you put on the hotter you feel. In the same way the thicker the blanket of greenhouse gasses around the earth the hotter Earth (or you and I) might feel (if you can prove at all)?

Therefore, the bottom-line, the "Climate Change" is being imposed from 'outside' not 'inside' the blanket. So, CO2 has no direct impact. Sure, CO2 would cause other adverse effects to the life on the earth- that is a separate issue.

Please feel free to keep in touch if I can assist.
Maheswar

A20
From: Mikki
Sent: Wednesday, September 11, 2013 9:38 AM
To: dkeefer@ciw.edu; Mike MacCracken (mmaccrac@comcast.net)
Cc: reeves@lanl.gov
Subject: panel for blanket?

09/11/13

Dear Dr. Meserve, President of Carnegie:
Dear Dr. MacCracken, Chief Scientist Climate Institute:

I was at the Panel discussion last evening. I could not stay until the end (because I had to be at another place)- I wanted to greet you, but missed.

Listening to the talk, I did not see much that I can contribute- being a new kid to the topic. And the panel thinking is very fluid- at least, I did not hear a workable concept or a plan?
As I said before- if any of you (or all of you) believe CO2 is the primary cause for the 'rising temperature', the best thing to do is- control emission of CO2- not putting more chemicals or dirt clouds (as a 'blanket') in space.
That is too risky and life threatening or worse than CO2 at 400ppm or 800ppm?

Now, getting back to the Reality:
I can show you the primary reason for this 'rise in temperature': The space in between our-Sun and the companion-Star is getting 'hotter' (for ~500 years) and 'colder' (for ~500 years) in 1,000 year cycles. And as the earth moves thru that 'hotter' space we feel 'rising temperature, melting ice, rising Sea levels, and destruction'.

We can't build an air-conditioner to cool the space, impossible. Nor can we build a 'blanket' around the earth and isolate it from the 'hotter' space??

It is like an Eskimo dressing in a wool-blanket in the polar region arriving suddenly into the equator region and refusing to give up the wool-blanket.

What would be the fate of this Eskimo at the equator?? Death.

That's exactly what would happen to the life on Earth if you place a wrong-blanket around the earth (if you can do that at all).

You are free to ask me- where is this companion-Star? How come our great Astronomers did not see it?? Or even if the companion exists how is the space in between gets 'hotter'???

Good Qs. That is my responsibility to explain to your satisfaction- I agree.

For now, I want to leave you with one more thought: our-Earth continues to travel in this ever increasing 'hotter space' for ~200 years more- then, the space begins to get 'colder' during the next ~500 years as our-Sun moves away from the companion-Star. And, this 1,000 year cycle appears ~700 times (Cycles) in the future before the End of The Time.

So, even if we are able to develop a safe blanket to cover Earth for about 250 years (in a 1,000 year cycle to protect life from getting destroyed by the hot-age), I am sure, only a handful can survive, like Adam and Eve at the end of 700 Cycles??

Please allow me to 'meditate' and devise a concept for a 'blanket' around the earth that would work in 'hot age' as well as in 'ice age'?

And, come to think of it "Almighty-Brahman" has already built that in place and it is working- can Man create a better blanket than the one built by "god"??

356

Ask NASA and Dr. Reeves at Los Alamos… may be they can help??
Please keep in touch if I can assist.
Thanks,
Maheswar

A21

From: Mikki
Sent: Friday, September 13, 2013 5:30 AM
To: Krimigis, Tom (Tom.Krimigis@jhuapl.edu)
Cc: reeves@lanl.gov; 'Edward Stone'
Subject: god-Isvar

09/13/13

Dear Dr. Krimigis:

*I sent you a message last evening and in today's MSNBC news
article I read your comments on Voyager 1 which said:*

[Surprises lie ahead
Even Tom Krimigis, a member of the Voyager team from Johns
Hopkins University, acknowledges that some of the puzzle pieces
still don't fit into place: For example, why hasn't there been a change
in the orientation of the magnetic field through which Voyager is
traveling? And why hasn't the cosmic ray flux evened out in all
directions? Both those phenomena were expected to occur when
Voyager crossed over into interstellar space, but neither of them has
happened yet.
"Initially we thought things were simple," Krimigis told NBC News.
"We'd cross the border, and we thought it was going to be flat and
boring, and we could all go home. But we found out that things
were anything but simple. I can tell you there are still some surprises
coming that we haven't published yet."
None of us will be alive when Voyager 1 breaks through the far side
of the Oort Cloud — but even though the spacecraft is a long way
from the limit of the solar system, Krimigis says Voyager's transition
is still well worth celebrating.

"It's like the <u>first Sputnik</u> that went beyond Earth's atmosphere, 56 years ago," he said. "That was a historic moment. This is another historic moment, going from the atmosphere of the sun to the atmosphere of the galaxy. The only difference is the altitude: 600 miles for Sputnik, as opposed to an altitude of 11.3 billion miles."]

I tend to agree with your thinking that "things are simple... but things were anything but simple"- you know why?? We are confused and we have no concept of the Reality.
Please look at the cover page of my Book which will be available next week where you can find answers to the "things" or most of the things we want to know- what is the Reality?

I wonder what other data would "tell...some surprises.."? I know for sure V1 or V2 will never "break through Oort Cloud" because it doesn't exist.
In fact, as I pointed out to you and Dr. Stone (couple of years ago) the V1 and V2 did not arrive at the Nose at all, but headed towards the Tail end of heliosphere- because, that's the way "things" work in Reality.

And your V1 at ~125 AU is not going anywhere beyond our-universe which is an Atom of ~722 AU diameter- it will be either destroyed at some point (if you can keep in touch with it you will find out) or reach our-companion Star to get burned? We will never know.

And that 125 AU distance is NOT a straight line measure- it is a distance on a curved path (an ellipse?).
V1 and V2 will keep going... going until hit something (another object in space) or get destroyed.

In fact your 125 AU could be wrong i.e. if you are calculating the distance based on the speed of light being constant at 3x10^8 m/s-

The speed of light at such great distances away from Sun would be less, many times less. If I know how you have calculated 125 AU, I can figure out the actual distance (on a circular path)??

I am counting on you and Dr. Reeves at Los Alamos do the right thing.
Please keep in touch.
Maheswar

A22

From: Mikki
Sent: Tuesday, September 24, 2013 11:18 AM
To: robert.gordon@heritage.org
Cc: 'jbast@heartland.org'; Al Gore <algore@algore.com> (algore@algore.com); IPCC-Sec@wmo.int; nicholas.culp@mail.house.gov; robert.hankins@mail.house.gov; chayva.lehrman@mail.house.gov; xan.fishman@mail.house.gov; karen_murphy@mikulski.senate.gov; royce_brooks@cardin.senate.gov; Kleinman, Joan <Joan.Kleinman@mail.house.gov> (Joan.Kleinman@mail.house.gov); carla.coleman@mail.house.gov
Subject: Nature v. CO2

09/24/13

Dear Mr. Gordon:

Thank you for listening to me, and thanks for forwarding my message to the scientists in your office. I look forward to meeting you and your experts, soon.

And, I know we can put this issue to bed forever.
I have the proof, convincing proof, to show why the 'rise or fall of temperature on the earth in 1,000 year cycles'?

I know, and I agree with you CO2 is not the cause, it is the nature. Just to say 'it is the nature' is not good enough- like the 97% of the scientists keep saying CO2 is the cause. They are confused and lost. Now, I am convinced (after listening to Dr. Carter & Dr. Soon- who are confused, too) that all 100% of the scientists are confused and lost.
Only point Dr. Soon made was "sun's radiation is the cause as a result of vast variations in the orbit-distances of Earth during thousands of years"- that is simply FALSE.

NASA came into existence ~50 years ago. We only have precise data on Earth's orbit since that time.

Sure, Newton had approximate data ~300 years ago.

And, no one- I say NONE- had any data for thousands of years based on which Dr. Soon can make such WILD arguments- Dr. Soon is totally out of reality.

I do know, around 1978-80, a scientist did publish a Curve, a Strange curve, based on some "strange" orbital data for the earth which he calculated from his "Fictional" equations- there is no proof that the earth actually took that kind of a WILD ride, like a drunk.

I only wish the leaders- you, Bast, Al Gore and Dr. Pachauri meditate deeply to figure out the real cause rather than going in different directions?

When I noticed this mess I decided to assist you all to come to a sensible understanding- because, I can convince you all the cause is the nature and why??

If I can convince you, I am sure we can convince the Congress and President to go in the right path....

What do you think? Are you willing to give me a chance to show you the PROOF that you need to move-on in the right-path??

Please see a copy of recent message to the members of the Congress!

Maheswar

From: Mikki
Sent: Thursday, September 19, 2013 10:51 AM
To: 'nicholas.culp@mail.house.gov'; 'robert.hankins@mail.house.gov'; 'allison.busbee@mail.house.gov'; 'chayva.lehrman@mail.house.gov'
Cc: xan.fishman@mail.house.gov; karen_murphy@mikulski.senate.

gov; royce_brooks@cardin.senate.gov; Kleinman, Joan <Joan. Kleinman@mail.house.gov> (Joan.Kleinman@mail.house.gov); carla.coleman@mail.house.gov; Al Gore <algore@algore.com> (algore@algore.com); IPCC-Sec@wmo.int; lars.heikensten@nobel. se; overbye@nytimes.com; Mike MacCracken (mmaccrac@comcast. net); FN-WHO-AAPI (AAPI@who.eop.gov); kevin_lefeber@ durbin.senate.gov; nicholas_devereux@warner.senate.gov

Subject: CO2 v. Rising temperature

09/19/13

Dear Congressman Upton, Whitfield and Waxman: [copy to Sen. Nelson, McCain and Whitehouse]

Yesterday, I was at your Committee 'hearing' on Climate Change and learned on the 'gist' of your concern. I know we are confused and divided. I hope I can help.

First, I want to thank 'Al Gore' and the 'Climate Reality' for informing me of this hearing timely. And, I thank our-great Senator Bill Nelson for sending me the details on this hearing. I admire his diligent work to the Nation.

Second, I want to thank Chairman Upton for meeting me at the hearing and accepting that 'one-page' (the cover page of the Book, see the attachment) I handed-over to him.
Third, I want to thank the 'staff assistants' of Chairman Whitfield and Ranking Member (or minority leader) Waxman for listening to me, accepting copies of that one-page, agreeing to circulate my message to all 31-Congressman in the Committee, and agreeing to give my message to Congressman Whitfield and Waxman.

So, what is my Message?

My message is simple.

The primary cause for the 'rise in temperature on Earth' is Nature, not CO_2. Sure, CO_2 could be detrimental for health/life of the species on Earth.

The phenomena of a mini 'hot-age' or 'ice-age' occurs in 1,000 year cycles [the Hot-age and Ice-age occurs in 4.32×10^6 year Cycles: please read my Book, "god-Isvar"]

Please allow me to explain on my message- see the one-page I attached [I will be happy to make a presentation with slides whenever you want me to do so].

As our-Sun (moving faster) approaches the companion-Star (moving slower), the space between Sun-Star heats up: during this time (~500 years) we will feel increasing temperature on Earth.

That means the temperature could keep increasing for another ~200 years until Sun-Star line up (see Figure) and we feel a mini hot-age (this is my estimate, being an old retired engineer).

Then, as our-Sun moves forward due to its higher speed, the space between Sun-Star begins to cool: during that time (~500 years) we feel decreasing temperature on Earth.

That meant we feel a mini ice-age in about ~700 years from now [of course, none of us would be here in the present form- although most of us will be back in another form or body since life, also, goes in cycles].

You may wonder where is this companion-Star?
My answer is simple.

All Astronomers, 100% of them on Earth, agree the space consists of 'double' Stars, everywhere as far as they can see, with 2-exceptions: (1) our-Sun is 'alone', and (2) there is a 'three Star' system next door.

The 100% of the Astronomers are wrong or confused to lump our companion-Star with a 'double-Star' next door and call it a three-Star system.

It sounds like Earth is the center of the universe- until Galileo declared not Earth, but Sun is the center. I know, not Earth or Sun, but Neutron is the center.

Similarly, the 97% of the scientists, who advocate CO_2 is the primary cause for the Climate Change or the rising temperature, are wrong or confused with the Reality. And, the primary cause is the Reality of Nature.

You, as a leader, should find out what is the Reality. To help you in that process may I suggest 2-simple steps:

(1) Allow me to make a presentation in the US Capitol to the members of the Congress and the staff.

(2) Call for a 'hearing' where I and other engineers/scientists (including those 97%) can testify on their findings- of course, the 97% of the scientists can review my Book (published this week) and come prepared to take me apart- I welcome the challenge.

Thereafter, we can put our minds to help solve the problem at hand, if there is a solution at all.

Thanks,

Maheswar

————————-

From: Mikki
Sent: Monday, September 23, 2013 6:39 PM
To: 'jlakely@heartland.org'; 'robert.gordon@heritage.org'
Subject: Nature v. CO2 Rising temperature

09/23/13

I am glad I came, listened, met and spoke to Mr. Bast and Mr. Gordon.
Mr. Bast asked me to come to Chicago- I said 'yes'. If he is going to be here, in Washington, for a day or two (I am sure it is possible)- I would love to meet Mr. Bast for few minutes (preferably with Mr. Gordon).
Later, I can come and visit with him in Chicago.

I am sure, you can guess- I am not impressed with Dr. Carter or Dr. Soon. They only spoke Negative- nothing Positive. In fact, they admit lack of knowledge as to the cause of rising temperature.
So, because of their own confusion, they simply decided to "deny" rising temperature. That is not professional, nor responsible to do.
I do not think they even took my challenge seriously that "I can show you the temperature is going to keep rising next ~200 years and I know the cause" (despite my agreement that CO_2 is not the cause).
If anyone of them is really serious, other than speaking negatively- they could have given me their card and asked to get in touch to see what 'hell' I have to say.

Therefore, I am counting on Mr. Bast, who seem to be a true leader, and Mr. Gordon, who has been involved in some aspects of the environmental issues.

I hope I can count on these two young leaders (of course, I am old) to sit down with me and see what I have to show. Please see the attachment.

I can help you put this divided issue behind us and use our-energy to do constructive work.

Maheswar

From: Mikki
Sent: Friday, September 27, 2013 8:43 AM
To: IPCC-Sec@wmo.int
Cc: Al Gore <algore@algore.com> (algore@algore.com); 'jbast@ heartland.org'; 'robert.gordon@heritage.org'; 'lars.heikensten@ nobel.se'; 'SpeakerBoehner@mail.house.gov' (SpeakerBoehner@ mail.house.gov); 'kevin_lefeber@durbin.senate.gov'; 'nicholas_ devereux@warner.senate.gov'; 'nicholas.culp@mail.house.gov'; 'robert.hankins@mail.house.gov'; 'chayva.lehrman@mail.house. gov'; 'karen_murphy@mikulski.senate.gov'; 'xan.fishman@mail. house.gov'; Kleinman, Joan <Joan.Kleinman@mail.house.gov> (Joan. Kleinman@mail.house.gov); Kleinman, Joan <Joan.Kleinman@mail. house.gov> (Joan.Kleinman@mail.house.gov); 'carla.coleman@ mail.house.gov'; 'royce_brooks@cardin.senate.gov'; 'bferguson@ sppinstitute.org'; Clinton Foundation (enews@clintonfoundation.org)
Subject: Nature v. CO2 why rise or fall of temperature?

09/27/13

Dear Dr. Pachauri:

[Copy sent to Whitehouse in response to President Obama's letter to me of 09/26/13]

I agree with your panel's finding that "...*the atmosphere and ocean have warmed, the amount of snow and ice has diminished, the global mean sea level has risen and the concentrations of greenhouse gases have increased...*"

But, I do not agree with your panel's conclusion that "...*man-made warming* (is) *extremely likely...*"

Yes, a small % of greenhouse gases are being made by man (the Reds in China, mostly- plus small pockets here and there??) which could act like a thin-blanket over the earth- if your scientists can show this thin-blanket can retain a small % of Earth's radiation from escaping into the space- similar to a condition when you put on a wool-blanket cover over your body at a time the outside temperature is <30F, that might do it. And if your conclusion that 'man made warming extremely likely' is based on that principle, you should ask your volunteer scientists to go back to the drawing board and measure <u>in situ temperatures below and above</u> the 'greenhouse cover'—then, only then, you can prove your conclusion is correct, if the temperature above the 'greenhouse cover' is <u>less</u> than that below.

Or if you find that the temperature <u>above</u> the 'greenhouse cover' is <u>higher</u> (or equal) to the value <u>below</u>, please let me know promptly— then, I can show you why?? In fact, I can assure you—the temperature <u>above</u> is going to be <u>high</u>—that meant the 'greenhouse cover' serves no real purpose in the long run... since the higher temperature <u>above</u> can get thru the 'greenhouse cover'. That is the Reality.

Your scientists can see the Reality in the Figure attached—if they review it carefully. The sun and its companion are making the space in between hot as they orbit (on X) and come closer during the next ~200 years (it is a part of 1,000 year cycle).

It is up to you, the experts- whether to see it or stay confused (or lost). I am willing to meet and explain to your experts on any day/time (in Washington or NYC) if it interests you—no, I do not force you to drink it.

Thanks for your recent acknowledgment of my prior request to 'delete 95% from your report'. I hope you will consider my suggestions and help put this dividing-issue to Bed: whether Nature or the insignificant Man is the cause for 95%?? If I do not hear from you promptly—I rest my case.

Thanks,
Maheswar
ps: by the way if you want to read my Book: god-Isvar, Swastika-science, please go to Amazon.com, click Books, and search for 'god-Isvar'. A revised version of the Book will be available in couple of weeks from now.

From: Mikki
Sent: Sunday, September 29, 2013 10:25 PM
To: IPCC-Sec@wmo.int
Cc: bferguson@sppinstitute.org; jbast@heartland.org; robert. gordon@heritage.org; Al Gore <algore@algore.com> (algore@algore.com); 'SpeakerBoehner@mail.house.gov' (SpeakerBoehner@mail.house.gov); FN-WHO-AAPI (AAPI@who.eop.gov); lars. heikensten@nobel.se; kevin_lefeber@durbin.senate.gov
Subject: Nature v. man's CO2 ??

09/29/13

Dear Chairman Pachauri:
[Copy to Mr. Obama in response to a message of 09/26/13]

Please refer to my message of 09/27/13, a copy of which you can see, below:
I am sorry, I did not provide you with a clear explanation why Nature is the cause at 95% certainty, not the man or the CO_2.

Actually, your use of 'man' is inappropriate because it is the scientists who created the CO2 mess with Nobel Prizes in hand- so, you should blame that 0.01% of the scientists, not the 'man'.

The issue is what next?? You and the rest of the scientists should attempt to understand the Reality- that is the next step.

Please, let us resist the temptation to start a 'new religion' in the name of CO2, like the confused-Pharisee did in the name of 'false-gods' before, during, and after the Persian Empire or the 'no-god' religion of the scientists.

Like I suggested before, I will be pleased to meet with any scientist who is willing to listen and learn about the Reality.

The ball is in your court to play the game- I know, we are going to lose if you play the same game of the confused scientists- I am putting it bluntly to get attention of the 'gang' which is leading the 'man' into a deep 'black-Hole', Einstein had created to 'beat' our 'godly' ancients- after studying Sanskrit writings… That is the kind of scientists I encounter all the time!

Please let me know if you and your team is different… If not, 'the next 3 or 4 generations will suffer the consequences- because the old people like me will be gone.

Thanks,

Maheswar

A23

From: Mikki
Sent: Tuesday, October 08, 2013 10:18 AM
To: lars.heikensten@nobel.se; 'p.w.higgs@ed.ac.uk'; 'interface@ulb.ac.be'
Cc: lawrence.krauss@asu.edu; overbye@nytimes.com; Al Gore <algore@algore.com> (algore@algore.com); IPCC-Sec@wmo.int; FN-WHO-AAPI (AAPI@who.eop.gov); ksamy@ceq.eop.gov; bferguson@sppinstitute.org; jbast@heartland.org; robert.gordon@heritage.org; 'incandel@hep.ucsb.edu'; Reid, Chip <ReidC@cbsnews.com> (ReidC@cbsnews.com)
Subject: god-particle v. godless-particle?

10/08/13

Dear Dr. Heikensten, Prof. Higgs and Prof. Englert:

First, I congratulate the Professors for theorizing, seeing, and winning the Prize.

Second, I hear Nobel F accepting CERN's god particle (or shall we say 'godless particle' as Prof. Krauss says).
In fact, Nobel F deserves a Prize for recognizing the black-magic show pulled out by CERN at a cost of Billions (or Trillion) of dollars- for how long?

It is said to be for quite a long time: $1/(1\times10^6)$ sec. of $(1\times10^9$ of 1×10^9 sec.)-time enough to get from here to heaven or hell- by the way, they do exist (read my Book).

What a great adventure, indeed a thrilling experience for all humanity to find that '2-protons' colliding at the speed of light create '125xproton mass' and disappear (into hell or heaven)- How long time, again??

I am impressed, just having heard of the News about this Magic. I am sure, Moses, the great Magician of all time would be impressed, too???

Please, for the 'god-particle' sake why not we STOP the drama of giving out Prizes for the Magic in Physics for some time, say- until after we come out of the present danger during the next 250-years??

What danger?? I am sure, you know what I mean- 'rise in temperature on the earth' due to CO_2.
Remember, CO_2 got the Prizes for doing the Magic in Physics or shall we say 'Peace in Physics'??

Also, please ask your Nobel experts in Physics, Chemistry, Medicine etc... etc... to review my Book: god-Isvar, Swastika-science (available at Amazon.com, Books... a revised version will be available in <two weeks).
And, come at me and try to take me apart- it is a challenge?? See if anyone is willing and ready to show me the 'god-particle or the 'godless-particle' of expert Kraus.

I am willing and ready to meet them on TV/Radio/Newspapers or in any arena they pick- is it fair??

By the way, I truly like your yesterday's award of the Prize in Medicine to three Professors- although they have no clue who is behind in creating the 'proteins' in the cell or transporting to the right place at the right time etc...

As they may be aware DNA, RNA or genes have no ability to do that.
If these good-Professors as well as rest of the Professors including those 100% confused-Scientists (especially those involved in CO_2 fiasco) would review my Book- they may discover the Truth. Or

Find out from me by trying to take me apart??

I really feel BAD to see my Brothers going in the wrong-Path and taking all humanity with them into a black-Hole created by the great Einstein and the likes in beating our 'godly' ancients (simply because their skin color happens to be different? Read the Book, you will get the whole story.)

Please feel free to write to me if I can assist.
Maheswar

A24

From: Mikki
Sent: Friday, October 25, 2013 12:29 PM
To: charles.bolden@nasa.gov; lars.heikensten@nobel.se
Cc: 'SpeakerBoehner@mail.house.gov' (SpeakerBoehner@mail.
house.gov); nicholas_devereux@warner.senate.gov; kevin_lefeber@
durbin.senate.gov; karen_murphy@mikulski.senate.gov
Subject: 'Newly detected galaxy is the most distant ever confirmed'-
Really?

10/25/13

Copy sent to <u>Senator Nelson</u>: This is the present status of our false-
research.
We are chasing distant galaxies which are <u>running away from us at
the speed of light</u> spitting Nobel.
We are chasing the "god-particle" when we crash 2 protons to get a
<u>fake 125 proton masses</u> spitting Nobel.
Of course, we have been chasing <u>black-Holes for 100 years</u> which
eats Billions and spits Nobel.
Let us not forget big-Bang and now the CO2 religion, they too spit
Nobel… where are we going with Nobel?

It seems we lost our senses and give Billions or Trillions to keep our
genius scientists keep doing false-Research while the Nobel keeps
spitting Prizes.
Is there anyone to look into this Mess??
Or the Drama is Great??

Maheswar

_____

From: Mikki
Sent: Friday, October 25, 2013 9:50 AM
To: 'stevenf@astro.as.utexas.edu'; 'papovich@physics.tamu.edu';
'tilvi@physics.tamu.edu'; 'aloeb@cfa.harvard.edu'
Cc: health-science@washpost.com
Subject: Reader feedback for 'Newly detected galaxy is the most
distant ever confirmed'

10/25/13

First, I want to thank writer Meeri Kim for putting the article together very well- so, even a young 7th grader can understand (of course I am old 7th grader trying to understand our genius scientists).

By the way, I am an old retired engineer in DC- I am involved in doing work from the point where 'Kepler' and 'Newton' have left us nearly 400 years ago.

I find we are in a deep black-Hole created by great Einstein and I am not sure how long it might take us to be out of that darkness?

Second, I want to thank the named scientists for their work in seeing, yet, another distant galaxy- I am not sure if the big-Bang scientists would ever understand the Reality?

Please allow me to put it in a 7th grader terms: we agree everything we see or sense is made of atoms, including you or I or the tree- correct?

And, the atoms make up a cell within you or I or the tree- correct?

So, what if I can show you "the earth-moon, planets-moons-ring objects, sun-star, plus neutron" make ONE ATOM in which we live- would you be curious enough to see what this old 7th grader is talking about?

Although you are sure that can't be true, if you agree to look at it, you may be surprised to find out the Reality.

That gives us a chance to start thinking in the same wave length—the big-Bang, black-Hole, expanding Universe or even the CO2 religion of 'rising temperature' etc.. etc.. will be gone with the wind?

And, galaxy is a cell of some ONE- who that might be?? If so, what are we? What are we doing confused about everything??

Please note: There is a flaw in the concept that "the free electron and free proton... later recombine to form neutral hydrogen again– however this hydrogen is in an excited state when formed, and as it relaxes back to the ground state, it emits a series of line photons."
See the reference paragraph below:
[If the Lyman Alpha Galaxy is defined as Young galaxies in the early universe can often be identified by prominent emission of Lyman alpha photons. The Lyman alpha line is the 2-1 transition of neutral hydrogen. Galaxies with vigorous ongoing star formation may be expected to show strong Lyman alpha emission lines, because they contain young, massive, hot stars along with neutral hydrogen gas. The hot stars emit copious amounts of ultraviolet radiation, which ionizes the neutral hydrogen, breaking it into a free electron and free proton. These particles later recombine to form neutral hydrogen again. However this hydrogen is in an excited state when formed, and as it relaxes back to the ground state, it emits a series of line photons. 2/3 of the time, this series ends with emission of a Lyman alpha photon. Lyman alpha emission was first suggested as a signpost of galaxy formation in by Partridge and Peebles in 1967.]

Also, please note: 'neutron' is the creator of proton, electron (or sun, planets...you or I or the tree); and it is the neutron that transmits the sound, color, light etc... in different frequencies which purely vary or depend up on 'entropy' (cold v. heat) at each point in space. That makes the speed of sound vary at a different rate, and the speed of light vary at a different rate from point to point in space- nothing is constant. So, the distance calculations based on red-shift or blue-shift can be erroneous. By the way, it is incorrect to say sun light takes 8 minutes to get to the earth- it takes much less time because the speed of light at the sun is many times faster than on the earth.

If you want to know more- please write to me or read my Book: "god-Isvar" (will be out end of this month: check Amazon.com, Books).

Thanks,

Maheswar

Newly detected galaxy is the most distant ever confirmed

From: Mikki
Sent: Tuesday, November 05, 2013 7:36 PM
To: overbye@nytimes.com; 'fdrake@seti.org'
Cc: 'epetigura@berkeley.edu'; 'ldoyle@seti.org'; charles.bolden@
nasa.gov; lars.heikensten@nobel.se; 'pnas@nas.edu'
Subject: Here comes Nobel?

11/05/13

Dear Friend Overbye and Dr. Drake:

First, Dr. Doyle and I were in touch in 2011 (and before) on a
topic of my work (which he could not understand) and on his work
"Kepler-16: A Transiting Circumbinary Planet" (Science, Vol. 333
no.6049 pp.1602-6, published on September 16, 2011) or "detection
of a planet whose orbit surrounds a pair of low-mass stars". I
understood what he has reported and gave him my findings based on
his own data. I made that a part of my Book: 'god-Isvar' (Amazon.
com, Books). I am planning to include a copy of this e-mail as a part
of that Book in a revised version which will be out by the end of this
month.

Second, I understand you are hunting for 'Earth-like' in our galaxy
and think "there could be as many as 40 billion habitable Earth-size
planets in the galaxy... (or there could be from less than a thousand
to a billion other civilizations in the galaxy)". Thank 'god' the
estimate came down to " a subset of some 42,000 brighter and well-
behaved stars.. (and found) 603 planets, of which 10 were between
one Earth and two Earths in diameter, and circle in... the habitable
zone, ... receiv(ing) between a quarter of the light the Earth gets,
and four times as much. In the solar system, that zone would spread
from inside the orbit of Venus to just outside the orbit of Mars...

(and) one out of every five sunlike stars in the galaxy has a planet the size of Earth circling it in the Goldilocks zone — not too hot, not too cold — where surface temperatures should be compatible with liquid water, according to a herculean three-year calculation based on data from the Kepler spacecraft".

[This is all based on Drake equation that consists of seven factors, which range over all human knowledge and aspiration (what does that mean, I am lost?)- O here is the answer:
"Some are strictly astronomical, like the rate at which stars are born in the Milky Way and the fraction of those stars with planets — 10 per year and half, respectively... Others are impossibly mystical, like the average lifetime of a technological civilization — 1,000 years to 100 million years was the guess. In between are more squishy details like how many habitable planets there are per planetary system (one to five), and what fraction of those habitable planets develop life, intelligence and the technology to communicate with other worlds... Multiply all the factors together and you get the putative galactic census. In the realms in which astronomers have actually gotten new data, the old guesses... have held up very well", ...the SETI institute, explained].

Third, now comes the red-meat finding: The above "result builds on a report earlier this year by ... the Harvard-Smithsonian Center for Astrophysics, who found that about 15 percent of the smaller and more numerous stars known as red dwarfs have Earth-like planets in their habitable zones... using slightly less conservative assumptions (Ravi) of Pennsylvania State University found that half of all red dwarfs have such planets".

You know, I can list more... what is the use? Who can argue... billions come your way and you get the free-Press, and of course Nobel is on the way!

Please allow me to report, again-

Third, you are wrong. There are NO smaller or bigger stars, nor red-dwarfs: it is the Neutron that obstructing your clear view; and, all stars should be of the same size as our-sun with planets. I made that point very clear to Dr. Doyle, NASA, Nobel, even to Overbye... (god knows who-else). I can't help if you guys want to go on in the wrong path milking billions... and Nobel.

Second, you are wrong and Drake Eq. is meaningless. Please go back and read Sanskrit writings: I understand, there are only a total of 13-habitable planets in our-galaxy; and, Earth is in the middle. Those of us who do the 'righteous duty' (like Buddha's disciple Zeus informed us in 0CE) might get to the above planet (in a different form, not like the present form), and those who do the 'unrighteous duty' (I am sure, we all know what that is?) might get to the below planet (in a different form to undergo severe suffering, or even end up on Earth in a different form to suffer). I know years ago, I did not believe it, nor do I expect you to believe me- now, I am convinced of that truth based upon my own work.

First, please read what I sent to Dr. Doyle in 2011 or at least read the Book: god-Isvar, written to a 7th grader, and see if that makes any sense??

If not, feel free to get in touch with me.
Mikki

The New York Times

November 4, 2013
Far-Off Planets Like the Earth Dot the Galaxy

By DENNIS OVERBYE

The known odds of something — or someone — living far, far away from Earth improved beyond astronomers' boldest dreams on Monday.

Astronomers reported that there could be as many as 40 billion habitable Earth-size planets in the galaxy, based on a new analysis of data from NASA's Kepler spacecraft.

One out of every five sunlike stars in the galaxy has a planet the size of Earth circling it in the Goldilocks zone — not too hot, not too cold — where surface temperatures should be compatible with liquid water, according to a herculean three-year calculation based on data from the Kepler spacecraft by Erik Petigura, a graduate student at the University of California, Berkeley.

Mr. Petigura's analysis represents a major step toward the main goal of the Kepler mission, which was to measure what fraction of sunlike stars in the galaxy have Earth-size planets. Sometimes called eta-Earth, it is an important factor in the so-called Drake equation used to estimate the number of intelligent civilizations in the universe. Mr. Petigura's paper, published Monday in the journal Proceedings of the National Academy of Sciences, puts another smiley face on a cosmos that has gotten increasingly friendly and fecund-looking over the last 20 years.

"It seems that the universe produces plentiful real estate for life that somehow resembles life on Earth," Mr. Petigura said.

Over the last two decades, astronomers have logged more than 1,000 planets around other stars, so-called exoplanets, and Kepler, in its four years of life before being derailed by a mechanical pointing malfunction last winter, has compiled a list of some 3,500 more candidates. The new result could steer plans in the next few years and decades to find a twin of the Earth — Earth 2.0, in the argot — that is close enough to here to study.

The nearest such planet might be only 12 light-years away. "Such a star would be visible to the naked eye," Mr. Petigura said.

His result builds on a report earlier this year by David Charbonneau and Courtney Dressing of the Harvard-Smithsonian Center for Astrophysics, who found that about 15 percent of the smaller and more numerous stars known as red dwarfs have Earth-like planets in their habitable zones. Using slightly less conservative assumptions, Ravi Kopparapu of Pennsylvania State University found that half of all red dwarfs have such planets.

Galaxy contains billions of potentially habitable planets, say University of California at Berkeley and University of Hawaii at Manoa astronomers.
Geoffrey Marcy of the University of California, Berkeley, who supervised Mr. Petigura's research and was a co-author of the paper along with Andrew Howard of the University of Hawaii, said: "This is the most important work I've ever been involved with. This is it. Are there inhabitable Earths out there?"

"I'm feeling a little tingly," he said.

At a news conference Friday discussing the results, astronomers erupted in praise of the Kepler mission and its team. Natalie Batalha, a Kepler leader from the NASA Ames Research Center, described the project and its members as "the best of humanity rising to the occasion."

According to Mr. Petigura's new calculation, the fraction of stars with Earth-like planets is 22 percent, plus or minus 8 percent, depending on exactly how you define the habitable zone.

There are several caveats. Although these planets are Earth-size, nobody knows what their masses are and thus whether they are rocky like the Earth, or balls of ice or gas, let alone whether anything can, or does — or ever will — live on them.

There is reason to believe, from recent observations of other worlds, however, that at least some Earth-size planets, if not all of them, are indeed rocky. Last week, two groups of astronomers announced that an Earth-size planet named Kepler 78b that orbits its sun in 8.5 hours has the same density as the Earth, though it is too hot to support life.

"Nature," as Mr. Petigura put it, "knows how to make rocky Earth-size planets."

Also, the number is more uncertain than it might have been because Kepler's pointing system failed before it could complete its prime survey. As a result, Mr. Petigura and his colleagues had to extrapolate from planets slightly larger than Earth and with slightly smaller, tighter orbits. For the purposes of his analysis "Earth-size" was anything from one to two times the diameter of the Earth, and Earth-like orbits were between 400 and 200 days.

Dr. Batalha said, "We don't yet have any planet candidates that are exact analogues of the Earth in terms of size, orbit or star type."

Though Kepler itself is sidelined while astronomers devise a new program it can accomplish with less flexible pointing ability, it has sent back so much data that there is still a whole year's worth of results left to analyze, Dr. Batalha said. "Scientists," she said, "are going to work on Kepler data for decades." Kepler was launched in 2009 to perform a kind of cosmic census, monitoring the brightness of 150,000 far-off stars in the Cygnus and Lyra constellations, looking for dips in brightness when planets pass in front of them.

Mr. Petigura and his colleagues restricted themselves to a subset of some 42,000 brighter and well-behaved stars. They found 603 planets, of which 10 were between one Earth and two Earths in diameter, and circled in what Mr. Petigura defined as the habitable zone, where they would receive between a quarter of the light the

Earth gets, and four times as much. In the solar system, that zone would spread from inside the orbit of Venus to just outside the orbit of Mars.

Meanwhile, in an innovation borrowed from other data-intensive fields like particle physics, Mr. Petigura designed a computer pipeline so that he could inject fake planets into the data — 40,000 in all — and see how efficiently his program could detect planets of different sizes and orbits. In addition to that correction, he and his colleagues had to correct for geometry; only about one in 100 planet systems is aligned edge-on so that earthlings would see the telltale wink of an exoplanet transit.

"It was a ton of work," he recalled, explaining that he had to try out tens of billions of different periods for each star in order to find planets.

Sara Seager, an exoplanet astronomer at the Massachusetts Institute of Technology who was not involved in the work, said the pipeline testing had made the results believable. "I would say that small planets are everywhere and very common," she said, "no matter how you slice and dice the data. But Kepler is dead and we have no way to get any further data. So we'll have to be satisfied with this as the final word, for now."

March 2, 2008
Ideas & Trends
Please Call Earth. We Still Haven't Found You.

By <u>DENNIS OVERBYE</u>

NEARLY half a century ago, Frank Drake, a young radio astronomer with extraterrestrials on his mind, stepped up to a blackboard in Green Bank, W.Va., and scribbled a string of symbols intended to bring some clarity to the question of just how alone humanity is in the cosmos.

The dozen wise men (there were no women) in the room were an elite group. Among them were Carl Sagan of Cornell University, as yet relatively unknown; the biochemist Melvin Calvin, who would learn during the meeting that he had won the Nobel Prize in chemistry; Barney Oliver, the research chief of Hewlett-Packard; and John Lilly, the dolphin expert, in whose honor the group dubbed themselves the Order of the Dolphin.

They sifted the variables in the light of what was then known or guessed, did the math, and concluded that there could be from less than a thousand to a billion other civilizations in the galaxy.

The Drake Equation, as it is known, has served as the bones of the search for extraterrestrial intelligence (SETI) and for the hopeful field of astrobiology ever since.

Since that meeting, in 1961, spacecraft have surveyed all the major bodies of the solar system, except for Pluto, and radio astronomers have listened for intelligent signals from more than 1,000 stars, so far in vain. Last month, a scaled-down version of our own solar system, with a pair of planets analogous to Jupiter and Saturn, was found orbiting a star 5,000 light years away in the constellation Scorpius, bringing the total number of known exoplanets, as they are called, to more than 250.

You might think we have made some headway in solving the equation, or rewriting it, or generally getting a handle on our cosmic loneliness. But you would be wrong. Astronomers today are as fuzzily optimistic (or pessimistic) as the Green Bank group.

"I get that question all the time," said Dr. Drake, 76, by phone from his office at the SETI Institute in Mountain View, Calif., where he is chairman emeritus and director of the Carl Sagan Center for the Study of Life in the Universe. "There hasn't been any great change. The equation still stands."

The discoveries of the last half-century, he explained, have confirmed what were just educated guesses on the part of the Dolphins.

Dr. Drake's equation consisted, and still consists, of seven factors, which range over all human knowledge and aspiration. Some are strictly astronomical, like the rate at which stars are born in the Milky Way and the fraction of those stars with planets — 10 per year and half, respectively according to the Dolphins. Others are impossibly mystical, like the average lifetime of a technological civilization — 1,000 years to 100 million years was the guess.

In between are more squishy details like how many habitable planets there are per planetary system (one to five they said), and what fraction of those habitable planets develop life, intelligence and the technology to communicate with other worlds. The Dolphins pegged those last three probabilities optimistically as, respectively, 100 percent, 100 percent, and 10 percent to 100 percent (dolphins, for example, don't build radio telescopes). Multiply all the factors together and you get the putative galactic census.

In the realms in which astronomers have actually gotten new data, the old guesses of the Dolphins have held up very well, Seth Shostak, an astronomer and spokesman at the institute, explained. In the more sociological and biological realms, where the data are ambiguous or nonexistent, you can't prove they were wrong.

"These guys were either enormously lucky or amazingly prescient," he said. One change, he said, was in the notion of a habitable world, that is to say, one with liquid water. In the old days "habitability" meant a planet had to be small and rocky and in a narrow Goldilocks zone around its star where the temperature would be just right.

Astronomers say space missions like NASA's Kepler, scheduled to be launched next year, will determine the frequency with which these Goldilocks planets occur in our neck of the galaxy. But the possibilities have expanded since spacecraft discovered evidence of water on or in some of the moons of Jupiter.

At the same time, scientists discovered that life on Earth was tougher and more versatile than scientists had thought, thriving in weird places like boiling undersea vents. "There is so much evidence for lots of pathways to the origin of life," Dr. Drake said.

But how often does intelligent and technological life actually emerge from such environments? Some evolutionists, like Stephen Jay Gould, who died in 2002, have argued that intelligence is not inevitable. The dinosaurs did just fine for 150 million years without getting appreciably brainier.

The advantages intelligence and technology confer, moreover, might also be outweighed by their dangers; thus the interest in the last term of the equation, the lifetime of a civilization. As Dr. Sagan would emphasize, this is the ringer in the works. Or, in Dr. Drake's words, "the real iffy one."

If the answer is less than a million years, said Geoffrey Marcy, an ace planet-hunting astronomer at the University of California at Berkeley, "Our Milky Way may be lit up only here and there during the past 10 billion years, and we may be the only lit bulb."

Dr. Sagan looked forward to finding an extraterrestrial signal as a sign that technological societies are not doomed to blow themselves up or poison themselves.

Once upon a time, the end of the cold war would have been seen as good news in that regard, but the possible advent of bioterrorism, genetical engineering mishaps and novel pandemics has led some otherwise sober thinkers, like Martin Rees, the Cambridge University cosmologist, to suggest that Civilization as We Know It won't make it out of the present century.

So the cosmic glass is still half full or half empty, depending on your personal inclinations. You can make your own guesses online, at msnbc.com/modules/drake/default.asp.

Dr. Drake, who conducted the first fruitless SETI listening tour, of a pair of stars in 1960, once said that the most likely aliens to hear from would be a race of immortals, who had plenty of time to wait for an answer. But he said now that he no longer expected to hear from ET in his lifetime. Under realistic estimates, he said, you would need to look at 10 million stars (there are 200 billion in the galaxy), and there is not enough time left.

"We could be wrong," he said. "The extraterrestrials have to help us a lot," he explained, by beaming powerful beacons our way.

But the trend might be in the opposite direction, if humans are any indication, he noted. Earth first became detectable in the 1950s, he said, when the planet was full of powerful television and radar transmitters beaming and leaking gigawatts of power into space.

"We assumed that was the way it was always going to be," both for us and, by extension, for extraterrestrials, he said. But now the big transmitters are being phased out in favor of cable and satellites that leak hardly anything at all out to space. It's very economical and it's the wave of the future. Earth is gradually going radio quiet.

"That's big change nobody anticipated," he said. Once the big powerful transmitters go off the air, he said, "We will still exist but we will be hard to detect."

From: Mikki
Sent: Thursday, November 07, 2013 11:54 PM
To: joel.achenbach@washpost.com
Cc: 'Petrus.M.Jenniskens@nasa.gov'
Subject: 'Asteroids may be colliding with Earth more frequently than scientists had thought'

11/07/13

Dear writer Joel:

Thanks for your article and the graph 'Path of the Chelyabinsk asteroid' by Patterson Clark. I consider the graph, assuming it displaying the correct-data, is an experiment by the Nature to show us that my ten-years of research in Cosmos is correct. I will add this e-mail and your article with the graph in the Appendix of my Book: 'god-Isvar' (revised version which may be out by the end of this month). In fact, this morning I wrote to 'Petrus.M.Jenniskens' on the very same issue- if he can provide me with such a graph or data…

Maheswar

Asteroids may be colliding with Earth more frequently than scientists had thought

Washington Post: By <u>Joel Achenbach</u>, Published: November 6, 2013

There are scads of building-size, potentially hazardous asteroids lurking in Earth's immediate neighborhood, and they may be colliding with the planet 10 times as often as scientists have previously believed, according to new research published Wednesday that examined the <u>airburst of a 25-million-pound asteroid</u> this year near the Russian city of Chelyabinsk.

Three studies released Wednesday, two in the journal Nature and one in the journal Science, provide the most detailed description and analysis of the dramatic event on the morning of Feb. 15.

To study the meteorite fragment and its internal components (including mineral composition), a 0.53g fragment of the Chelyabinsk meteor, which exploded over Russia on Feb. 15, 2013, was imaged using X-ray tomography. This work was done by Douglas Rowland (Center for Molecular and Genomic Imaging) and Qing-Zhu Yin (Department of Earth and Planetary Sciences) at University of California, Davis.

Graphic: see Fig.9.1e

Scientists now estimate the diameter of the object at just a hair under 20 meters, or about 65 feet. Undetected by astronomers, the rock came out of the glare of the sun and hit the atmosphere at 43,000 mph.

As it descended through the atmosphere, it broke into fragments, creating a series of explosions with the combined energy of about 500 kilotons of TNT, making it more than 30 times as powerful as the atom bomb that destroyed Hiroshima in 1945, although the energy in this case was spread out over a much broader area.

The shock wave blew out windows in nearly half the buildings in Chelyabinsk. It knocked people off their feet; dozens were sunburned by the blinding flash, which at its peak was 30 times as bright as the sun. About 1,200 people were hurt, most by broken and flying glass, but no one was killed.

One chunk the size of a love seat landed in frozen Chebarkul Lake and left a cosmic bullet hole in the ice. That fragment, which weighed about 1,900 pounds, was retrieved months later, breaking

390

into several pieces in the process. Thousands of smaller pieces have also been recovered.

The scientific investigation relied to a great degree on video imagery obtained by "dash cams," the cameras Russian drivers often use to document car crashes and potentially abusive law enforcement officers. Scientists visited 10 locations where the footage had been taken by stationary cameras and used landmarks to map the asteroid's trajectory. The damage of the shock wave propagated perpendicularly to the rock's path.

"It's incredible how well documented all this is," said Peter Jenniskens, who is a meteor astronomer at the SETI Institute and a co-author of the paper in Science.

Taken together, the new information on Chelyabinsk does not suggest that the sky is falling (no one has ever been killed by an asteroid in all of recorded human history).

But it may shift the overall risk profile of asteroids, making Chelyabinsk-size events look more probable.

That's the conclusion of a new Nature report, "A 500-kiloton airburst over Chel¬yabinsk and an enhanced hazard from small impactors."

The scientific orthodoxy held that a Chelyabinsk-size event ought to happen roughly once a century. But Peter Brown, a professor at Western University in London, Ontario, and his collaborators reexamined decades of data compiled by government sensors. They saw several events that seemed to defy the odds.

Famously, a very large object exploded over the Tunguska region of Siberia in 1908. There have been less-heralded impacts, including one on Aug. 3, 1963, when an asteroid created a powerful airburst off the coast of South Africa.

"Any one of these taken separately, I think you can dismiss as a one-off. But now, when we look at it as a whole, over a hundred years, we see these large impactors more frequently than we would expect," Brown said.

NASA's asteroid experts said they weren't ready to recalibrate their estimates of impact frequency just yet. Paul Chodas, a scientist at NASA's Jet Propulsion Laboratory who works on near-Earth objects, noted that on the same day that the Chelyabinsk asteroid hit, another asteroid — completely unrelated to the first — nearly hit the planet, too.

"We had two very rare events the same day, and immediately one can start asking questions about whether these are as rare as we thought," Chodas said at a news conference Wednesday. "But I would point out that these are still small-number statistics."

His colleague Don Yeomans, also of the Jet Propulsion Laboratory, echoed that point: "I would be more comfortable if that conclusion were backed up with more data."

Most rocks in the Chelyabinsk range have yet to be identified, and it would be difficult and expensive to find them and calculate their trajectories, Brown said. But this could spur the development of new space telescopes and other early-warning systems, he said.

The paper in Science hypothesized that the Chelyabinsk asteroid is a piece of "rubble" from a larger body that had been broken apart by tidal forces from an earlier near-Earth encounter.

"The rest of that rubble could still be part of the near-Earth object population," the authors wrote.

Author

The author *Maheswar*, a retired engineer of 75 years old, is born in *Krishna* river delta, Andhra Pradesh, India—close to the birth place of *Nagarjuna*, a teacher of righteous duty in the East as a disciple of *Buddha*, like Jesus in the middle-East and in the West. The author has been involved in the research of this topic for more than ten-years as a resident in Washington, DC, United States of America. This research of space began at a point where Man is left by Kepler and Newton nearly 400 years ago- wrongly assuming Sun is the center of our-universe. Sun is not, but the creator of the sun, earth-moon and the planets-moons, the man-animals-birds, and the rest on the earth—*Indra* is the center of our universe.